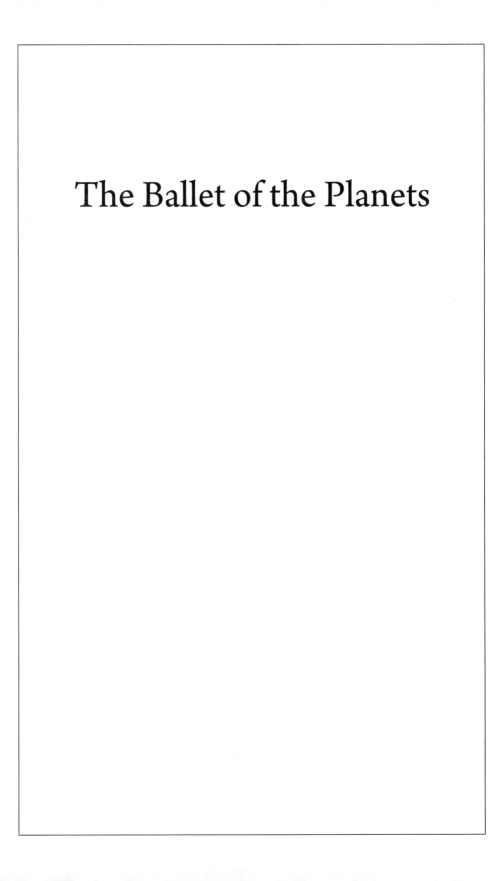

The Ballet of the Planets

The Ballet of the Planets

On the Mathematical Elegance

of Planetary Motion

Donald C. Benson

OXFORD

UNIVERSITY PRESS

OXFORD
UNIVERSITY PRESS

Oxford University Press

Oxford University Press, Inc., publishes works that further
Oxford University's objective of excellence
in research, scholarship, and education.

Oxford   New York

Auckland   Cape Town   Dar es Salaam   Hong Kong   Karachi
Kuala Lumpur   Madrid   Melbourne   Mexico City   Nairobi
New Delhi   Shanghai   Taipei   Toronto

With offices in

Argentina   Austria   Brazil   Chile   Czech Republic   France   Greece
Guatemala   Hungary   Italy   Japan   Poland   Portugal   Singapore
South Korea   Switzerland   Thailand   Turkey   Ukraine   Vietnam

Published by Oxford University Press, Inc.
198 Madison Avenue, New York, New York 10016

www.oup.com

Library of Congress Cataloging-in-Publication Data
Benson, Donald C.
The ballet of the planets : on the mathematical elegance of planetary motion / Donald C. Benson.
p.   cm.
Includes bibliographical references and index.
ISBN 978-0-19-989100-9 (hardcover : alk. paper) 1. Planetary theory—History. I. Title.
QB361.B46   2012
523.2—dc23   2011033632

135798642
Printed in the United States of America
on acid-free paper

Dedicated to my wife, Dorothy, without whose help and
loving encouragement this book would not be possible

CONTENTS

ACKNOWLEDGMENTS

I would like to thank the following people for reading the entire manuscript and making many important suggestions: Dorothy Benson, Ned Black, and Donald Chakerian. I also thank my anonymous reviewers for their invaluable suggestions.

INTRODUCTION

...the celestial ballet is beheld

in its repeated performances

before our eyes.

Nicholas Copernicus, *De revolutionibus* (1543)

And first of all, astronomy ... taught that there are laws.... these rules were discerned by Hipparchus, Ptolemy, Copernicus, one after the other, and finally, ... Newton ... Henri Poincaré (1854–1912) *The Value of Science* (1907)

I grew up in a Los Angeles neighborhood from which the Griffith Observatory is visible, high on a hill to the north. I am sure that its presence had a subtle influence, turning my young mind toward science and mathematics. Eventually, mathematics became my passion and my profession. In this book—returning to the source of my inspiration of long ago—I will tell the story of an ancient astronomical problem, the problem of understanding planetary motion. Such understanding means giving the motion of the planets a mathematical structure.

I believe that this book would have interested me when I was a high school student who loved mathematics. I hope that it will be of interest to all—including students, professionals, and hobbyists—who enjoy mathematics.

Early astronomy, the topic of this book, constitutes the birth of science and the forerunner of the technology that is such an important part of modern life— ending the struggle for mere survival and giving you and me the time to read and write books. The motion of the planets was very puzzling to early scientists because, over a period of weeks, the positions of the planets relative to the fixed stars changed in a manner that was difficult to explain. This book looks at several early theories of planetary motion—illustrating the development of mathematical models in the exact sciences. The first geometrical theories came from Greek astronomers in the fourth century BCE (although the Babylonians had earlier learned to predict planetary phenomena by using repeating arithmetic patterns). This book concludes with Isaac Newton's treatment of this problem in the seventeenth century, which, even today, is sufficient for the needs of the space program.

Planetary astronomy, like science in general, advanced through the interplay of observation and theory. Theory gives an abstract explanation of observation, and observation provides a critique that confirms, weakens, or denies the theory. Even without the stimulus of new observations, an old theory can be replaced by a new one that explains the old observations in a simpler or more general manner. The heliocentric theory of Copernicus is an example of a simpler new theory superseding a more complicated old one.

Although this book contains historical facts—and I want them to be accurate—I cannot claim that this is a history of planetary astronomy, for the following two reasons: first, I have not attempted to give a *complete* history of the subject. I have included some historical topics that I believe are relevant to the development of mathematical models of planetary motion. Second, I wish to discuss the ideas of astronomers of the past—in particular, how their ideas lead to the science of today—but I do not feel the need to use their original presentations. In this book, I have selected from the history of astronomy a few ideas, mainly geometrical, that lead forward to the current science of the solar system. My choice of historical topics is eclectic and personal. I do not claim to give a *complete* history of planetary astronomy.

In a 1676 letter to Robert Hooke, Isaac Newton (1642–1727) wrote, "If I have seen further it is only by standing on the shoulders of giants." In this book, I discuss—from the point of view of current, not ancient, science—a few of the contributions of these giants, including Archimedes, Ptolemy, Copernicus, and Kepler, as well as Newton himself. I have selected contributions that led forward and made possible the science of today. The scientific methods and standards of the ancient scientists differed from those of today, but I leave that topic to the historians of science.

Naked-eye astronomy is uniquely suitable as an introduction to the exact sciences because there are no scientific prerequisites. The theoretical side of positional astronomy involves nothing deeper than elementary geometry. On the other hand, electromagnetism, quantum mechanics, and relativity involve deep concepts that are so far beyond ordinary experience that these topics are inappropriate as an introduction to the exact sciences for the nonspecialist. Furthermore, current concepts of cosmology—for example, string theory—remain largely speculative, untested (perhaps untestable) by observation. Books on science for a general reader often tell *what* is true, not *why* it is true. Because the basic concepts in this book are not difficult, it will be possible to tell not only what is true but also why.

The development of the study of planetary astronomy cannot be understood without also telling the story—beginning with Archimedes (287–212 BCE)—of the parallel development of mechanics, the branch of physics dealing with the behavior of physical bodies subject to movement or force. These two threads of knowledge, astronomy and mechanics, become beautifully entwined in Newton's theory of universal gravitation.

Geometry plays the central role in all theories of planetary motion. Geometrical concepts are self-contained in this book, but I do assume a reader will have a background of high school algebra. Two geometrical curves are fundamental to the development of planetary astronomy: first epicyclic curves and later ellipses. I have devoted separate chapters to the mathematical properties of these curves.

The telescope was invented shortly before planetary motion was shown to be elliptical. Nevertheless, the telescope did not play a significant role in the discovery of elliptical orbits or the law of universal gravitation. This book is concerned with a science based on naked-eye observation in a simpler time when the planets were mere points of light and five in number: Mercury, Venus, Mars, Jupiter, and Saturn.

The observational data for positional astronomy, the measurements of the positions of celestial bodies, were obtained long before the invention of the telescope by the use of instruments for measuring angles—for example, a special purpose protractor called a *quadrant* (Figure 2.6 on page 20). In contrast, modern scientific instruments, such as the cyclotron, require an esoteric chain of logic to "see" the result of an experiment.

Science progresses from the interplay between observation and experiment on the one hand and theory on the other. In the study of the movement of the planets, as in the exact sciences generally, a theory often takes the form of a mathematical structure, a *mathematical model*. This book deals with the mathematical models that were inspired by the movement of the planets, from Ptolemy's epicyclic curves to Newton's ellipses.

Science provides instruments of observation that supplement our senses. More important, science provides the thought patterns, often mathematical, that lead from speculation to validation. Science is nourished by cycles of belief and skepticism:

1. *The belief that the universe, however chaotic it may seem from day to day, is governed by unambiguous rules that can be understood and tested.* Science depends on the few who passionately pursue these theories. Science prefers theories of the greatest possible generality and simplicity.
2. *The skeptical pursuit of every speck of evidence—every observation, measurement, or calculation—that might show that a proposed rule is false.* The believer and the skeptic may or may not be the same person.

A set of rules, as in 1, is called a *scientific theory*. In ordinary usage, a theory can be tentative or fanciful—a "pet theory"—but in science *theory* has more serious overtones. A theory must be more than a guess. In science, a guess is often called a *conjecture* or a *speculation*.

If two theories explain the same phenomena, the simpler one is preferred. This principle is sometimes called *Occam's Razor* after the fourteenth-century English philosopher William of Occam (c. 1285–1349).

Although the *motivation* of scientists is varied and complex, the *validation* of a scientific theory is achieved through the cyclic alternation of propounding a theory, as in step 1, testing that theory, as in step 2, modifying the theory in the light of critical observation, and so on. Scientific validation provides an alternative to authority—the authority of a spokesman, a book, or a tradition.

Ancient positional astronomy put forward two principles:

1. All astronomical movement is relative to a motionless Earth. (This was the consensus view, despite the contrary suggestion of Aristarchus of Samos.)
2. All astronomical movement is somehow based on uniform circular rotation.

Starting in the sixteenth century, astronomy took three revolutionary advances—the subject of the later chapters of this book.

1. The Sun, not Earth, is the center of planetary movement.
2. Planetary motion is elliptical, not circular.
3. Gravity controls planetary motion.

The ballet of the planets, the debut of science, is performed in three acts. The first act is played under naive assumptions: Earth is the center of the universe, and all heavenly bodies follow paths based on uniform circular motion. But there are hints of disquieting problems.

The second act opens with brilliance and confusion. The dancers—the planets—have all changed their costumes. Old assumptions are discarded. Earth is no longer at the center, and circles are replaced with more complex curves.

In the third act, much (but not all) is explained. We finally learn what forces have drawn the dancers together.

The Ballet of the Planets

1

THE SURVIVAL OF THE VALID

It can scarcely be denied that the supreme goal of all theory is to make the irreducible basic elements as simple and as few as possible without having to surrender the adequate representation of a single datum of experience.
Albert Einstein, (1934) *Phil. Sci., 1*(2), 163–169.

This chapter examines the process of scientific validation. In the last hundred years, science has achieved giant steps forward. In this great success, scientists played three essential roles.

1. Researcher: The scientist makes discoveries and publishes them.
2. Mentor: The scientist prepares others to undertake scientific work.
3. Referee: The scientist judges the validity of the work of other scientists.

The ancient astronomers were not subject to rigorous refereeing. However, some of their offerings would stand up to the scrutiny of a hypothetical ancient referee and would lead in a direct line to the science of today. In this book, I have attempted to select a few such results in the field of planetary astronomy.

The path of scientific research is greatly influenced by the system of mentoring. The senior researcher conveys important information such as what problems are important and what methods are promising to a protégé. The Greek philosopher Plato (427?–347? BCE) was an important mentor of Greek astronomers; his influence extended far beyond his lifetime.

Plato

The puzzle of planetary motion prompted Plato to urge the astronomers of his day to "save the appearances." This cryptic remark is often interpreted to mean "find an explanation of the seemingly erratic motion of the planets," but scholars tell us that Plato intended a much narrower meaning: "Show that the motion of the planets can be explained in terms of *uniform circular motion*." Plato required that any explanation of planetary motion must somehow involve uniform circular motion because he claimed that the motion of celestial bodies must be perfect and that uniform circular motion is the only perfect motion.[1] Of course, Plato vastly underestimated the complexity of the universe. Indeed, the success of modern science is based on complex ideas and subtle experimentation. Nevertheless, Plato is expressing an important motivating principle for scientists—that the world is simpler than it seems, and every effort should be made to discover its simplicity.

Plato's insistence on uniform circular motion shows his prejudice in favor of circles and spheres, which were studied with such success in geometry. The importance of circles was also confirmed by the circular paths of the fixed stars. Plato saw perfection in these figures, and he exerted his great influence to limit the discussion of the motion of celestial bodies exclusively to uniform circular motion. In fact, this prejudice in favor of uniform circular motion persisted until early in the seventeenth century when Johannes Kepler (1571–1630) demonstrated that the planets follow *elliptical* orbits with *nonuniform* velocity.

Plato considered the realm of ideas more real and important than observations with the five senses. This was made clear in his *Allegory of the Cave*, which describes prisoners in the cave who are chained so that they see only shadows on the cave wall from a great light shining into the entrance of the cave. The shadows represent physical appearances and the great light is from the world of ideas. According to Plato, the shadows are much less important than the great light beyond. In Plato's opinion, the world of appearances, including any scientific observation, is less important than the realm of ideas.

Nevertheless, Plato's dictum to save the appearances is an important early step in the history of science. An important theme in this book is how the dictum of Plato to search for uniform circular motion encouraged initial progress in astronomy but, after a time, blocked new ideas.

Plato was the founder of the Academy in Athens. Currently, "the academy" sometimes refers to the worldwide network of higher learning. Science, including astronomy, is one branch of the academy.

The worldwide academy has a forum with all the excitement of a courtroom drama—the surprise witness, the crucial evidence, the convincing summation. However, unlike the courtroom, this forum lacks a definite location, a judge, and a jury of twelve. Nevertheless, the academy judges scholars and their ideas through the process of *peer review*.

1.1. PEER REVIEW

Peer review is a standardized procedure for the *evaluation* of scientific research. On the other hand, the *creation* of scientific research can be more chaotic. Style, authority, convention, and prejudice may enter, thereby enhancing the diversity of science. The rebel and the naysayer may discover errors in the previously accepted scientific canon, and science may benefit. Even if a scientist is inspired by seeing visions or hearing voices, peer review is the standard by which his or her work is judged.

When an author believes he has achieved some important original research, he writes an article and submits it to an academic journal dealing with his research topic. The editors of the journal then send the article to one or more *referees*—experts in the area of the author's research. A referee is ethically bound to review the work impartially and must refuse to review if she has any personal connection with the author or his work. The anonymous referee, her identity known only to the editors, volunteers her time to evaluate the submitted article's validity, significance, and originality.

The referee sends her detailed review to the editors with a recommendation: publish the article as written or with certain revisions, ask the author to revise and resubmit, or reject the article. The referee's comments are generally communicated anonymously to the author. Of course, the editors make the final decision to publish or not. Today, scientists are unlikely to accept a new result unless it has been published in a peer-reviewed journal.

The practice of rigorous peer review by professional journals dates roughly from the middle of the twentieth century. Thus, the astronomers discussed in this book were generally not subject to the process of peer review. We cannot judge their work by current standards, but we can judge their work's *validity, significance,* and *originality* relative to the scientific knowledge of that earlier time.

1.2. THE SCIENTIFIC METHOD

A referee judges the *validity* of a scientific article primarily on the basis of the *scientific method*—illustrated here by an early theory of the *movement of the fixed stars.*

The ancient Greeks generally believed that the regular nightly motion of the stars from east to west was explained by the following principle:

Hypothesis 1.1. *The stars are rigidly attached to a rotating celestial sphere concentric with the Earth.*

The concept of a scientific theory is central to the scientific method.

Definition 1.1. A *scientific theory* is a general principle or a set of principles that can explain or predict the outcome of empirical observations.

The following two steps are fundamental to the scientific method.

1. Observation of *data* in an area of interest. For example, the ancients observed that the stars form fixed patterns, or constellations, that move in unison from east to west. In some sciences, data can be acquired by means of experiment.

2. Construction of a *theory* to explain the data. For example, Hypothesis 1.1 is a possible explanation for the above observation.

The purpose of a theory is to organize and generalize the data. The theory must be capable of predicting further observations and is considered correct to the extent that these predictions are verified. For example, the theories of positional astronomy, also called *astrometry*, organize observations of the heavens and predict the movement of celestial bodies.

A scientific theory should strive for simplicity, avoiding concepts that are not related to the observed data. As we learned, this principle is known as *Occam's Razor*. Paraphrasing the epigraph from Einstein (1879–1955) at the beginning of this chapter results in a version of Occam's Razor that warns against oversimplicity.

Principle 1.1 (Occam's Razor). *Scientific theories should be as simple as possible* but no simpler.

The *scientific method* consists of an ongoing cyclic alternation of the above steps 1 and 2. In the cycles after the first, steps 1 and 2 are replaced with the following:

1a. Observation of data predicted by the theory. In this step, the theory is tested as rigorously as possible by data that might possibly contradict the theory. Confirmation of a scientific theory is always approximate because observation is always subject to error. The closer the agreement is between theory and observation, the more convincing the confirmation.

2a. Possible modification of the theory. If the theory is confirmed by observation, we look for even more demanding tests of the theory. Otherwise, if inconsistencies are discovered, the theory must be modified or discarded.

Thus, the scientific method is a two-step process, alternating between belief and skepticism. The scientist wears two different hats: one is of the theoretician and the other is of experimentalist. A consistent philosophy is elusive.

The theoretician is the Platonist. She seeks to *save the appearances*, explaining the facts with ideas of the greatest possible simplicity and generality. The great theoreticians—e. g., Copernicus, Newton, and Einstein—are the most exalted names, the heroes of science. The theoretician rejoices in unifying concepts. The skeptical experimentalist/observer is the anti-Platonist. He destroys weak and ineffectual theories.

A theory designed to explain some initial data is frequently called a *hypothesis*. A hypothesis in the absence of any confirming data—that is, steps 1 and 2 without any further cycles of the scientific method—is often called *speculation*. For example, the atomic theory of the ancient Greek philosopher Democritus (460?–370? BCE)—that the fine structure of matter consists of discrete indivisible particles called atoms—was speculation because it was not confirmed by data of any kind.

To confirm this theory, one would have to show that matter cannot be divided into arbitrarily small pieces. The Roman poet Lucretius claimed that the atomic theory was confirmed, for example, by the fact that the scent from an open perfume bottle reaches us from across the room without our being able to see what has moved from the bottle to us. This argument does not confirm the atomic theory because it does not deal with the question of unlimited divisibility.

In the nineteenth century, thousands of years after Democritus, the existence of atoms was finally confirmed by observation. On the other hand, Aristotle (384–322 BCE) speculated that in free fall a heavier object falls faster, but this was disproved centuries later by Galileo (1564–1642) who dropped heavy and light objects from the Leaning Tower of Pisa. Aristotle's speculation was proven false, but Democritus's was confirmed.

A science never consists of just one theory, but rather it is made of an interlocking web of theories and subtheories. A healthy science produces continuing cycles of the scientific method, especially as applied to the more dependent subtheories. For example, Copernicus asserted (1) the theory that the Sun is the center of the solar system, and he fleshed out this heliocentric theory with (2) a subtheory defining the orbits as particular epicyclic curves, curves used earlier by Ptolemy for this purpose. Theory (1) survived cycles of the scientific method, but subtheory (2) was destroyed when Kepler showed that the orbits were ellipses.

Mathematical theories have a different source of validation called the *axiomatic method*, which has its origin in ancient Greek geometry. Using the axiomatic method, theorems to be proved are linked, using *rules of inference*, to axioms and previously proved theorems. Thus, the correctness of a proposition is established by verifying the validity of logical inferences leading back to the axioms—initial assertions accepted without proof.

Before mathematical validation, there must come mathematical *discovery*, which is based on trial and error, analogy, intuition, and creativity.

Geometry

Geometry was a towering mathematical achievement of the ancient Greeks. The world's first geometer is said to have been Thales of Miletus (now Turkey) (624?–547? BCE). He is said to be the first to devise geometric proofs. A compilation of the full flowering of Greek geometry is contained in the *Elements* of Euclid (fl. 300 BCE).

This effort was so successful that the geometric propositions of Euclid are still studied today worldwide by millions of high school students. Although mathematicians may debate the foundations of Euclidean geometry, today, all of the propositions of Euclid are considered fundamentally correct.

Is geometry a science—that is, is geometry a study of the physical world? To a degree it is, because points, lines, and planes are mathematical models for objects that exist in the physical world. However, geometry ignores its physical roots once its axioms are admitted to the canon. The celebrated Dutch mathematician B. L. van der Waerden gave the title *Science Awakening* to his book on the early history of mathematics— mostly of geometry.[2] Nevertheless, geometry is a science like no other.

Modeling science on the pattern of geometry has a benefit and a danger.

The benefit is that geometry shows the power of *abstraction*. For example, in geometry a line, unlike the lines we might draw with pencil and a ruler, has no width. Yet the theorem that the sum of the angles of a triangle equals $180°$ can be verified to a high degree of accuracy by using the draftsman's tools. Furthermore, this abstraction gave geometry a conceptual beauty that it could never have achieved if lines came in various widths.

Science had to wait until the seventeenth century for the benefits of similar abstraction when Newton conceived of the useful fiction of point mass.

The danger in modeling science after geometry is that geometry has little place for observation and experimentation.

Ancient astronomy began with a theory that explained many observations. When further observations did not fit the theory, efforts were made to reconcile the new observations with the theory, and the theory was abandoned when it was impossible to establish sufficient agreement with observation. This process has often been repeated in the history of science.

This chapter discussed the scientific method—the process of validation in science. In the next chapter, we examine the beginnings of Greek astronomy, with the two-sphere model of the universe.

PART ONE

Birth

2

THE BOWL OF NIGHT

Awake! for Morning in the Bowl of Night
Has flung the Stone that puts the Stars to Flight:
And Lo! the Hunter of the East has caught
The Sultan's Turret in a Noose of Light.
Omar Khayyám (1048?–1131?) *Rubáiyát translation by Edward FitzGerald (1859)*

The tranquil perfection of a star-filled sky contrasts with the chaotic confusion of everyday life on the Earth. Thus, rational explanations of the physical world started with the heavens. From antiquity, the regular movements of the Sun, Moon, and stars have provided a calendar, a clock, and a means of navigation. The early Polynesian explorers settled the Pacific islands with little else to guide them.

The stars, Sun, Moon, and planets are excellent initial objects of study for ancient science because, unlike the economy or the weather, they exhibit *stability*, a quality difficult to find in the chaotic world around us. This concept of stability is embodied in the saying "As surely as day follows night."

Of course, we now know that the stability of the *fixed* stars is illusory. The stars and galaxies move with incredible speed. They only appear "fixed" because they are extremely far away. This fortunate circumstance provided a fixed platform from which to study the movement of closer celestial objects—the Sun, Moon, and planets.

The stars are mysterious because, although we can see them, they are too far away to touch. Where are the stars? What are they? From ancient times, legends sought to explain this mystery. Science first arose from the effort to find reliable explanations. Scientific thought gave us an eye into the universe long before the telescope.

The crystalline celestial spheres

The ancient Greek astronomer and mathematician Eudoxus of Cnidus (408–355 BCE) posited a system of impenetrable crystalline spherical shells centered about the Earth to house the heavenly bodies—one shell each for the Moon, the Sun, and the planets, and an outermost shell for all of the fixed stars. The solidity of these shells was never confirmed by any observation, yet this speculation was given some credence even by Copernicus, Kepler, and Newton. We will see later how the supposed impenetrability of these shells impeded the thinking of Ptolemy and others. However, in the next section we see that the outer shell containing the fixed stars was a useful construction.

2.1. THE TWO-SPHERE UNIVERSE

Aristotle describes an important early theory in astronomy, the *two-sphere universe*, consisting of the Earth and the celestial sphere.[1] This theory makes two claims, the first concerning the Earth:

Hypothesis 2.1 (The spherical Earth). *(a) the Earth is spherical. (b) It is stationary and occupies the center of the universe.*

A theory, as above, that places the Earth at the center is called *geocentric*. In contrast, a theory that places the Sun at the center of our planetary system—the *solar* system—is called *heliocentric*.

A second hypothesis concerns the *fixed stars*. The existence of a spinning Earth seems to me to be implicit in the term *fixed* stars, but this idea was rejected by Aristotle. These stars were called fixed because they seem to move rigidly, as a unit. They appear to keep the same distances one from another as they rotate about the polar axis. They provide a fixed background against which one can observe, as discussed in the later chapters of this book, the more erratic motion of other celestial bodies: the Sun, the Moon, but especially the planets.

Hypothesis 2.2 (The celestial sphere). *The fixed stars are attached to a much larger sphere, the celestial sphere, concentric with the Earth. This sphere rotates from east to west at a uniform rate about an axis through the north and south celestial poles.*

The ancient Greeks raised the spherical Earth (Hypothesis 2.1a) to a level beyond speculation because they not only described the theory but also provided observational support. Consider, for example, the following three observations:

1. When ships far at sea first sight land, they see only the high elevations. As they approach closer to shore, the lower elevations also appear.
2. As one travels south, the constellations in the southern sky rise higher above the horizon.
3. During a lunar eclipse, the shadow of the Earth is round.[2]

Aristotle also argued that the Earth is spherical because all terrestrial objects tend to move toward the Earth's center—a remarkable foretelling of one aspect of universal gravitation. However, if Aristotle's contemporaries judged scientific validity by today's standards, I believe they would consider Aristotle's argument an interesting, but overreaching, speculation and items 1–3 more convincing support of a spherical Earth.

Aristotle and almost all of the ancient Greek philosophers and mathematicians believed the Earth to be the center of the universe and motionless. As evidence for the motionlessness of the Earth, Aristotle offers only common sense. We have no feeling of the Earth's motion. If the Earth were rotating, a falling object would follow a path distorted by the Earth's rotation. Unfortunately, Aristotle had no way to quantify this effect because Newton's theory of dynamics came more than two thousand years later. As we now know, the path of a falling object *is* distorted by the Earth's rotation, but this effect is negligible in ordinary circumstances. However, see page 144.

Eighteen centuries before Copernicus, Aristarchus of Samos (310?–230 BCE) proposed the heliocentric theory in which the Earth spins about its polar axis and orbits about the Sun, but his suggestion did not gain acceptance.

The geocentric view of the universe is supported, not only by common sense, but also by the universal human feeling that we are the most important occupants of the universe. We have been disabused of this comfortable feeling because modern astronomy has shown that our Earth is a mere speck in an incredibly vast universe.

The theory of the motionless Earth is not strictly wrong. In a sense, it is an arbitrary choice to consider the Earth spinning or fixed. Like children on a merry-go-round, we can fancy that we are spinning and the rest of the world fixed—or that we are standing still and the world outside is rotating past us. The only fault with this is that it complicates further study. From a modern perspective, a stationary Earth makes it difficult to explain, for example, the behavior of a Foucault pendulum—a massive pendulum hung from a high ceiling, displayed in some science museums (for example, at the Griffith Observatory). Even though the pendulum begins swinging in a precisely vertical plane, the rotation of the Earth causes this plane to change—to precess—clockwise in the northern hemisphere and counterclockwise in the southern hemisphere.[3]

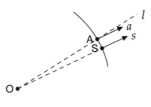

Figure 2.1

The ancient Greeks estimated the circumference of the spherical Earth. The mathematician Eratosthenes (276–195 BCE) attacked the problem as follows. He noted that in the Egyptian town of Syene (now Aswan) at the summer solstice the Sun at midday was directly overhead,[4] whereas on the same day in his hometown of Alexandria, approximately due north from Syene, the Sun at midday was located one fiftieth of a a circle (i. e., 7.2°), south of overhead. In Figure 2.1, Alexandria and Syene are the points A and S, respectively, and O is the center of the spherical Earth. The parallel arrows a and s are directed toward the Sun from these two locations. Eratosthenes observed that the angle between the arrow a and the line l was 7.2°, and he concluded that the angle AOS must also be 7.2° because a line through the center of a circle must meet the circumference at 90°. Because Eratosthenes knew that Alexandria was located at a distance 5,000 stadia north of Syene, he concluded that this distance must be $7.2/360 = 1/50$ of the entire circumference of the Earth, which, therefore, must be

$$5,000 \times 50 = 250,000 \text{ stadia}$$

The exact value of this unit of distance, the stadium, is not known, but it is believed that Eratosthenes' figure for the circumference of the Earth was between 39,690 km and 46,620 km. The true value is 40,008 km. This uncertainty is due partly to our ignorance of the true value of the stadium and partly to Eratosthenes' error in measuring distances and angles.

The celestial sphere

According to Omar, the stars are attached to an inverted "Bowl of Night." Indeed, an inverted hemisphere—for example, a planetarium ceiling— is a natural representation of the nighttime sky. The celestial sphere is the view of the sky seen by an observer at its center.

The celestial globe

The celestial globe, which represents the sky in the manner of a world globe, is a physical model of the celestial sphere.

Figure 2.3 depicts a hemisphere of a celestial globe showing the midnight sky on the vernal equinox at Beijing, Madrid, and Philadelphia—indeed, anywhere on the circle of latitude 40° North. The figure is a view of the outside of the celestial globe—an orthogonal projection such that the outer circumference represents the horizon on the vernal equinox at latitude 40° N. Like all of the geometrical figures in this book, Figure 2.3 was constructed with the graphics language MetaPost. For each of the hundreds of stars shown, this program reads data (right ascension, declination, and magnitude) from the *Bright Star Catalogue*.

It is easy (but, as we know now, wrong) to argue for the existence of an actual sphere concentric with the Earth containing the fixed stars because these stars circle in unison about the north celestial pole, continuously rising in the east and setting in the west.

This movement is remarkably regular.[5] Observing the night sky through the seasons, one obtains a map of the celestial sphere. Depending on whether the observer is north or south of the equator, there is a circular region centered, respectively, about the south or north celestial pole that can never be observed. On the equator, the entire celestial sphere can be observed in the course of a year.

The fixed stars are not "attached" to an actual celestial sphere because today we know that the fixed stars are located at a variety of distances. Hence, it would be incorrect to assign a particular radius to the celestial sphere. However, we can "save" the celestial sphere by redefining it in a way that avoids the problem that a sphere must have a particular radius. Define a "point" of the celestial sphere to be an infinite half-line emanating from the observer. The celestial sphere becomes a bundle of infinite half-lines emanating from the eye of the observer. This concept of celestial sphere expresses our actual observation of the line of sight to each star, but avoids any consideration of the distance to each star.

Thus, the celestial sphere formalizes our commonsense perception of the nighttime sky. It is the natural and correct concept for navigators who use the position of the stars to find their way; moreover, it is a basic concept used by all astronomers, ancient and modern.

Our view from *outside* the celestial globe is the mirror image of what the hypothetical observer *inside* the globe sees. If you can imagine that the observer writes "Help me" on the inside of the (transparent) globe, on the outside we will see "ɘm qlɘH."

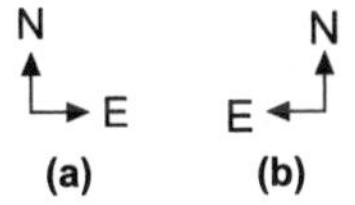

N ↑ └→ E   N E ←┘ ↑

(a)   (b)

Figure 2.2 (a) The terrestrial and (b) the celestial compasses.

Figure 2.4a is the usual *terrestrial* compass. If we paste a transparent decal of this compass on the outside of the celestial globe, then, if N points to the north celestial pole, E will point correctly eastward. However, our hypothetical observer inside the celestial globe will look up and see Ǝ ←┘ (N↑). As a matter of convenience, he might rewrite the letters И and Ǝ to obtain the usual *celestial* compass in Figure 2.4b.

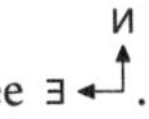

The celestial globe uses the geographical map-reading convention and the terrestrial compass (Figure 2.4a). However, the view looking up at the night sky is the mirror image of Figure 2.3—governed by the celestial compass in Figure 2.4.

Parallax

Parallax is a concept that enables us to measure the distance to remote in-accessible objects. The ancient Greeks understood this concept, but they did not possess the accurate instruments necessary to measure the distance of stars.

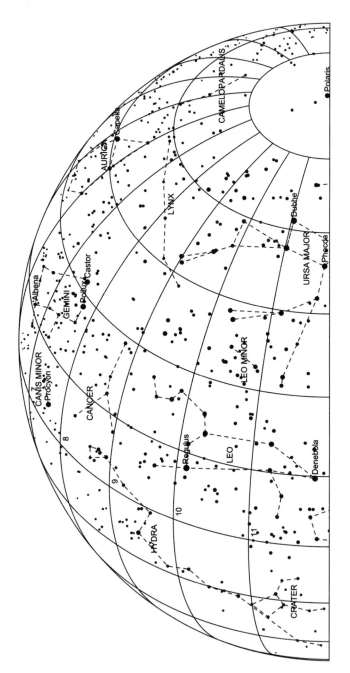

Figure 2.3 Orthogonal projection of a celestial globe. The midnight sky on the vernal equinox at latitude 40° North.

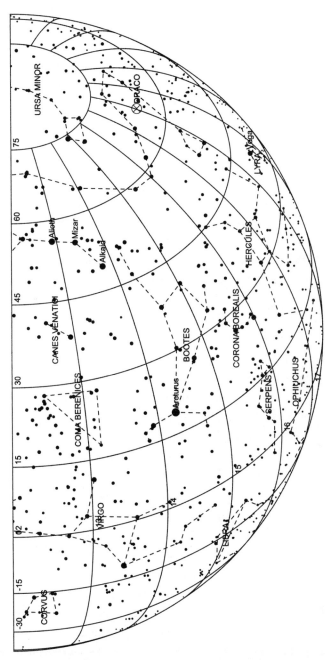

Figure 2.3 Continued. The symbol ⊗ in the constellation Draco denotes the *North Ecliptic Pole* (NEP), discussed in connection with Figure 2.8.

Figure 2.4 Parallax.

Parallax is the change in the direction to an object when viewed from different vantage points. This is illustrated in Figure 2.4. The object O is viewed from equidistant points P and Q. If the distance d is known, then the length of OP or OQ is a straightforward trigonometric computation. If the angle a is small and measured in radians, then the lengths OP are OQ are approximately equal to d/a.[6]

Ancient Greek astronomers noted that the constellations of stars do not exhibit parallax as the celestial sphere rotates and that they appear the same regardless of the point of observation. They offered this in support of the contention that the Earth is stationary at the center of the celestial sphere. In fact, parallax of the fixed stars was negligible because the distance to the nearest star is about 200,000 times the diameter of the Earth's orbit. At that time, the smallest angle detectable was about 10 arcminutes, but parallax of the nearest stars involves angles of 1 arcsecond or smaller.

The problem of *lack of parallax* had to wait almost two millennia for a satisfactory solution using precise means of astronomical observation. The parallax of a star (61 Cygni) was first observed in 1838 by the German astronomer Friedrich Wilhelm Bessel (1784–1846). To maximize parallax, two observations can be made at an interval of six months, so that the two sightings are from opposite points on the Earth's orbit around the Sun, a distance of about 300,000,000 kilometers. A parallax of 1 second of arc[7] observed at this distance defines 1 parsec, an astronomical unit of distance. One parsec is equal to about 3.26 light years. The distance to 61 Cygni, one of the closest stars, is about 3.5 parsecs, or about 11 light years. (There are just 14 stars in the Milky Way that are closer than 61 Cygni.)

Current technology is able to use parallax to track objects, not only in the solar system, but even far out in our galaxy. We will see later how, in the seventeenth century, Kepler devised a clever parallax method to track planets using a much weaker technology.

The ancient astronomers were unable to determine the absolute distances of the planets, but they came close to finding the relative distances. For example, Ptolemy knew that the number 1.52 was important in determining the motion of Mars, but he did not realize that this number was the ratio of two distances: the mean distances from the Sun of Mars and of Earth. Nicholaus Copernicus (1473–1543) was the first to interpret these numbers as relative distances. He did so without using any observations that were more precise than those of antiquity, and he had no need of the concept of parallax. He supported his solution of the relative distance problem with purely theoretical arguments in connection with his heliocentric theory. The astronomer Tycho Brahe (1546–1601) rejected Copernicus's heliocentric theory of the solar system but accepted his basic pattern of relative planetary distances.

The artificial satellite, Hipparcos, launched by the ESA (European Space Agency) in 1989, which used parallax to map objects in our Milky Way galaxy, measured angles in the range of thousandths of an arcsecond (milliarcseconds). Thus Hipparcos improved the accuracy of naked eye observations by a factor of about one million.

One arcminute[8] is roughly the portion of the sky that can be observed through the eye of a sewing needle held at arm's length. To approximate an arcsecond, the needle must be seen at a distance of about 150 feet, and for a milliarcsecond, about 30 miles.

Projected astrometric satellites, NASA's SIM (Space Interferometric Mission) and ESA's GAIA (Global Astrometric Interferometer for Astrophysics), will measure parallax with accuracy of a few microarcseconds. To approximate a microarcsecond, the eye of the needle must be seen from 30,000 miles.

Binocular vision, rangefinder cameras, and *stadia surveying instruments* all determine distance, in a manner similar to the above astronomical example, through the use of parallax.

Binocular vision in humans (and other creatures) uses the minute differences in the images seen by our two eyes to instruct the arm, hand, and fingers exactly where to grasp an apple. Thus, parallax enables the two-dimensional images from our eyes to inform us about the three-dimensional world in which we live.

A rangefinder camera determines the distance of an object by using images from two separate lenses. The photographer sees these two images through the viewfinder. When he superimposes the two images of the object by turning the focusing knob, a complex gear linkage focuses the camera lens on the object.

The stadia surveying technique requires two pieces of equipment: a stadia rod and a special telescope called a stadia transit. The stadia rod, about 2 meters in length, is simply a graduated measuring stick, as shown in Figure 2.5b. Looking through the transit, one sees one vertical and three horizontal crosshairs, as shown in Figure 2.5a. The upper and lower horizontal crosshairs, (a) and (c), called the *stadia hairs*, encompass a fixed angle of view through the transit. The distance between the stadia hairs is

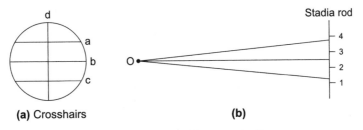

Figure 2.5 The stadia method for measuring distances.

arbitrary, but, for convenience, it can be set so that when the transit is horizontal—that is, when the transit and rod are at the same elevation—the length of the portion of a vertical object cut off by the stadia hairs is exactly 1/100 of the distance from the object to the transit. Thus, the transit operator is able to determine the distance to the vertically held stadia rod by reading the distance shown on the rod between the stadia hairs and multiplying by 100. The stadia method can be modified for the case that the transit and rod are at different elevations.

The definition of parallax applies to the stadia surveying method because the direction from points on the stadia rod to point O, the transit, changes when seen from two different points on the stadia rod. Thus the usual roles are reversed: the transit becomes the object observed and the stadia rod the observer.

2.2. COORDINATE SYSTEMS

Angles, not distances, are the fundamental astronomical measurements. Angles are used to define points on the celestial sphere by using the systems of coordinates discussed below.

The quadrant (Figure 2.6) was the basic instrument for measuring the angle between a star and the horizon. The astrolabe was an ancient instrument for the same purpose. The sextant, more precise than the quadrant, was used by navigators starting in the eighteenth century. The quadrant and its variants served as the instrument of choice for astronomers up to and including Tycho Brahe in the sixteenth century (Figure 8.1). It was the quadrant, not the telescope, that provided sufficient precision for Kepler to conceive his laws of planetary motion and thereby to open up the modern view of the solar system.

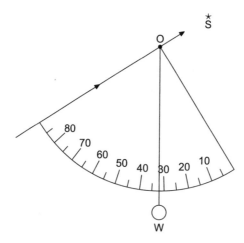

Figure 2.6 The quadrant. The device pivots at O. A weight at W is attached with a string to O. Sighting along the arrows determines that the angle of elevation of the star S is 32°.

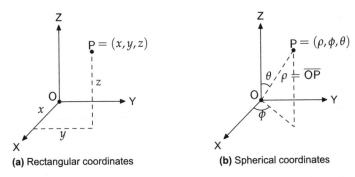

(a) Rectangular coordinates   **(b)** Spherical coordinates

Figure 2.7 Three dimensional coordinates of a point relative to three mutually orthogonal axes. The three spherical coordinates, ρ, ϕ, and θ, are called the radius, longitude, and colatitude, respectively.

A point in three dimensions can be specified mathematically by rectangular or spherical coordinates. Specifically, in Figure 2.7, the point P, with respect to the coordinate system XYZ can be specified by either of the triples (x, y, z) or (ρ, ϕ, θ). Spherical coordinates satisfy the following inequalities:

$$\rho \geq 0 \qquad 0° \leq \phi < 360° \qquad 0° \leq \theta \leq 180°$$

Taking ρ to be constant, (ϕ, θ), longitude and colatitude are coordinates specifying a point on the surface of a sphere. The angle θ is the angular distance to the z-axis—the north polar axis—and ϕ is the angle measured along the equator to the x-axis. Both the terrestrial and the celestial spheres are given coordinates in this manner, with minor differences. These coordinate systems are described below by relating them to the mathematical spherical coordinates. For the terrestrial sphere there is one system of coordinates, and for the celestial sphere there are three. These four coordinate systems are defined below by relating them to the mathematical definition in Figure 2.7b.

1. Terrestrial longitude and latitude.

Longitude is measured east and west of the polar meridian through Greenwich, England. Latitude is measured north and south of the equator. Terrestrial latitude is equal to 90° minus the colatitude (θ in Figure 2.7b), where positive and negative latitudes are interpreted as north and south, respectively. Terrestrial longitude and latitude are universally used to specify a point on the Earth's surface—for all map making and navigation.

2. Azimuth and elevation.

The z-axis passes through the zenith and the x-axis points to the horizon at due north. Azimuth and elevation are the coordinates for raw observations of a point in the sky. Azimuth is measured in degrees clockwise from north. Elevation is the angle above the horizon. The azimuth and elevation of a star (other than the North Star) changes moment by moment.

A quadrant (Figure 8.1) is used to measure the elevation. An *azimuth quadrant* is a revolving quadrant mounted on a pedestal with a scale measuring the azimuth of an observation. Ancient astronomical observation was done with an azimuth quadrant or similar instrument.

The azimuth and elevation of a star depend on the geographical location of the observation, and they change through the night as the celestial sphere slowly rotates westward about the polar axis. Astronomical observations of a celestial object need to be related to the movement of the celestial sphere, that is, to the westward rotation of the fixed stars. This could be done by observing the azimuth and elevation of the object and a known fixed star simultaneously, or as nearly simultaneously as possible. In the Renaissance or later, when accurate chronometers were available, it was sufficient to note the time of observation. The above method could be used to create a chart of the fixed stars or to log the movement of a planet.

3. *Right ascension and declination.*

A star chart requires a system of coordinates such that each star has a fixed pair of coordinates, independent of the time or place of observation. Either right ascension and declination or ecliptic longitude and latitude serve this purpose.

The z-axis passes through the north celestial pole and the x-axis passes through a point in the constellation Pisces, corresponding to the position of the Sun at the vernal equinox. Right ascension is usually measured in hours instead of degrees: 24 hours = 360°. Right ascension increases eastward. Declination is measured in degrees north and south like terrestrial longitude. Declination in degrees is labeled in 15° increments on the left side of Figure 2.3b. Right ascension in hours is labeled along the celestial equator in Figures 2.3a and b. Right ascension is similar to terrestrial longitude with the following differences:

1. Right ascension is measured in hours instead of degrees (1 hour = 15 degrees).
2. Right ascension takes values between 0 and 24 (including 0, excluding 24). Right ascensions from 0 to 12 and from 12 to 24 correspond, respectively, to longitudes *west* (from 0° to 180°) and *east* (from 180° to 0°).
3. The zenith at noon on the vernal equinox has right ascension 0. (The zenith at midnight, shown in Figure 2.3, on the vernal equinox has right ascension 12 hrs.)

A star, unlike a planet, is *motionless* in the sense that its right ascension and declination are unchanging.[9] Right ascension and declination are universally used for modern astronomical observations. Today, astronomical observatories are outfitted with a motor that moves the telescope, the observer, and the slit in the observatory dome so that, once the telescope is set to a certain right ascension and declination, it remains so directed until reset.

4. Ecliptic longitude and latitude.

In this system, the circular path of the Sun about the celestial sphere, the *ecliptic*, plays the role of the equator. Thus, the z-axis passes through the north ecliptic pole. Ecliptic longitude increases eastward. The x-axis, the same as for the previous system, passes through the position of the Sun at the vernal equinox. Ecliptic latitude is measured north or south along meridians orthogonal to the ecliptic. Ecliptic longitude and latitude were used to describe the positions of the planets because the planets follow paths through the zodiac close to the ecliptic circle. Ecliptic longitude and latitude are more stable than right ascension and declination because the ecliptic longitude and latitude of a fixed star does *not* change due to the 25,785-year precession of the equinoxes, a phenomenon discussed later in this chapter.

The last three coordinate systems have the following similarities:

- Elevation, declination, and ecliptic latitude are all measured in degrees north and south of the horizon, the equator, or the ecliptic, respectively.
- Azimuth, right ascension, and ecliptic longitude all increase in the eastward direction.

In addition to the quadrant (Figure 2.6), accurate astronomical observations require a clock. At nighttime, the movement of the fixed stars is an accurate timekeeper. In daytime, the Sun might be expected to play this role, but we will see that, without extensive correction not available to ancient astronomers, the Sun cannot be used to tell the time reliably.

2.3. THE SUN

The ancient Greeks believed that the Sun inhabited its own separate sphere and that there were many "crystalline" spheres that generated the "music of the spheres" while governing the movements of all celestial bodies including the stars, Sun, Moon, and planets. The significance of these spheres seems to be more poetic than scientific. It is possible to discuss the Greek scientific contributions without referring further to the crystalline spheres. A single *celestial sphere*, introduced earlier in this chapter, is sufficient.

The Sun moves relative to the celestial sphere. If this were not so, there would be stars too close to the Sun to ever be observed. In the daytime, no stars are visible because of the brightness of the Sun. Nevertheless, even in the daytime sky, constellations of stars are present because their locations have been charted when the Sun was elsewhere. Thus, we can say, for example, that the Sun is in the constellation of Leo although the stars of Leo are not visible.

The Sun follows an eastward path on the celestial sphere at a rate slightly less than one degree per day. This path, the *ecliptic circle* in Figure 2.8, is a great circle inclined 23.5° with respect to the celestial equator. Using the heliocentric premise, the plane containing the ecliptic circle, the *ecliptic plane*, is the plane containing the Earth's orbit around the Sun. The inclination of the ecliptic circle, the *obliquity* of the ecliptic, is equal to the 23.5° tilt of the Earth's axis of rotation with respect to any line perpendicular to the ecliptic plane, the plane containing the Sun and the Earth's orbit. In Figure 2.8, the arrow shows the eastward direction of movement of the Sun. The twelve constellations that meet the ecliptic are known as the *zodiac*: Aries, Taurus, Gemini, Cancer, Leo, Virgo, Libra, Scorpio, Sagittarius, Capricorn, Aquarius, and Pisces. The Sun makes a complete eastward circuit of the ecliptic each year.

The diameter of the celestial sphere perpendicular to the plane of the ecliptic intersects the celestial sphere in two points: the north and south ecliptic poles. The north ecliptic pole (NEP) is shown in Figure 2.8, and also in Figure 2.3 where it appears in the constellation Draco.

The NEP is a more stable point than the north celestial pole. To explain this, it is useful to revert to the modern heliocentric premise. The Earth slowly wobbles as it rotates. This wobbling motion, called *precession*, resembles the wobbling motion of a spinning top. Due to precession, the north celestial pole executes a circle about the NEP every 25,785 years. Thus, Polaris is only temporarily the North Star. In fact, in Ptolemy's time there was no bright star to indicate true north. The location of the Sun at the vernal equinox is shown at the left margin of Figure 2.8. Due to precession described above, the Sun at the vernal equinox moves gradually west from one constellation of the zodiac to the next. For this reason the precession of the Earth's axis of rotation is called the *precession of equinoxes*. The vernal equinox is currently moving from Pisces to Aquarius (hence, the song "The Age of Aquarius" from the 1968 musical *Hair*).

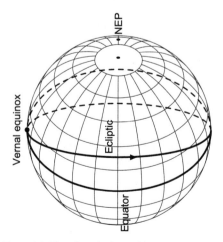

Figure 2.8 The celestial sphere, showing the equator, the ecliptic circle, and the north ecliptic pole (NEP).

The Sun is our basic timekeeper, dividing time into days and years. This seemingly simple concept is complicated by the fact, discussed below, that the apparent day is not constant.

Day by day

The *apparent day* is the period of time from one noon to the next, where noon is defined as the moment the Sun crosses the local meridian—the great circle on the celestial sphere that contains the north and south poles and the local zenith. Surprisingly, the apparent day is not constant—for two reasons:

1. The tilt of the Earth's axis of rotation.
2. The variation in the Earth's orbital speed. This fact was discovered by Kepler early in the seventeenth century, together with the fact that the orbit is an ellipse, not a circle.

The apparent day can gain or lose as much as 20 seconds per day, resulting in a cumulative error of as much as 15 minutes.

The period of time called the *mean solar day* is the average length of the apparent days. The mean solar day, divided into hours, minutes, and seconds, is the time standard for astronomical and ordinary daily needs throughout the world. Henceforth, the word *day* without any qualifier is an abbreviation of *mean solar day*.

A *sidereal day* is the time interval for any fixed star (other than the north star) to complete a circular path about the celestial pole—in other words, the time required to return to the same celestial meridian on two successive nights. The sidereal day is equal to 23.9344696 hours. This very stable value is the period of rotation of the Earth on its axis. The sidereal day is shorter than the mean solar day because in a day's time the Sun moves slightly eastward relative to the celestial sphere by about $1°$. Since the sidereal day is slightly shorter than the solar day, the constellations of stars march westward through the night sky as the year progresses.

Year by year

A *solar* or *tropical year* is defined as the time interval between two successive vernal equinoxes, currently a period of 365.2424 days. The solar year is the target of nearly all calendrical systems—for example, the Gregorian and Julian calendars—because the tropical year is defined by the recurrence of the seasons. The Julian calendar was discarded because the seasons gradually migrated through the calendar. A vernal equinox occurs at the instant that the Sun crosses the celestial equator from south to north.

The tropical year, divided into days, hours, and minutes, is the basis for the system of time measurement called *mean solar time* by which we set our watches and carry on our daily lives.

A *sidereal year* is defined as the length of time for the Sun to traverse the entire zodiac—that is, the time of one revolution of the Earth about the Sun—about 365.2465 days. The sidereal year is more stable than the tropical year.

The tropical year is different from the sidereal year because of the westward precession of the equinoxes. The lengths of the sidereal and tropical years and the phenomenon of precession were first measured by Hipparchus (190–120 BCE). He is said to have used data from the Babylonians to obtain these measurements. Remarkably, the two-sphere geocentric universe of Greek astronomy was adequate conceptual support for his calculations.

Hipparchus compiled a trigonometric table, the first of a genre that became indispensable, not only to astronomers, but also to engineers and physical scientists, until the invention of the hand-held calculator.

Hipparchus's improvement on Aristotle's model of solar motion

The following incident illustrates of the use of the scientific method in antiquity. Aristotle, together with many of his contemporaries, claimed that the Sun moved in a circular orbit at constant speed with the Earth at the center. Hipparchus discovered facts that were inconsistent with this model. He proposed an alternate model that matched the data more accurately.

This story, illustrated in Figure 2.9, foreshadows a hallmark of modern science—the use of *meticulous observation* to detect *subtle inconsistencies* with a theory. Hipparchus noted the unequal lengths of the seasons: northern hemisphere spring and summer are both slightly longer than either autumn or winter. Hipparchus observed that the unequal lengths of the seasons is inconsistent with Aristotle's claim that the Sun moves eastward in a circular path on the ecliptic with *constant speed*—a claim previously supported by the philosophical principle of the perfection of uniform circular motion.

Using current knowledge, unavailable to the ancient Greeks, the bold outer curves in Figures 2.9a and b represent the Earth's elliptical orbit about the Sun, or, equivalently, adopting the ancient geocentric premise, the Sun's orbit about the Earth. The outer tick marks represent the position of the Sun at equal time intervals. The tick marks at the top of the figure are slightly closer together than those toward the bottom.

In Figure 2.9a, the inner curve is a circle centered at the Earth E. The inner tick marks represent equal distances traveled in equal time intervals. The inner and outer tick marks do not line up exactly, showing a discrepancy with Aristotle's theory.

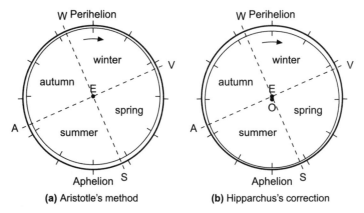

(a) Aristotle's method   **(b)** Hipparchus's correction

Figure 2.9 Hipparchus improves Aristotle's model of the Sun's motion. In both (a) and (b), the outer curve is the elliptical (nearly circular) clockwise motion of the Sun relative to the Earth, seen from south of the ecliptic plane. As discovered by Kepler early in the seventeenth century, this curve is an ellipse, currently with eccentricity 0.016710, with the Earth at a focus E of the ellipse. The outward tick marks are placed at twelve equal time intervals beginning at perihelion—the closest approach of the Earth and Sun, occurring in the first week of January. These marks coincide roughly with the Sun's annual visitation to each of the twelve constellations of the zodiac. W, A, S, and V denote, respectively, the *winter solstice, autumnal equinox, summer solstice,* and *vernal equinox.* The lines WS and AV intersect at E in a right angle. *Perihelion* and *aphelion* are the points at which the distance between the Earth and Sun are, respectively, smallest and greatest. The apparent speed of the Sun in its eastward journey is greater near perihelion and less near aphelion.

Aristotle believed that the Sun moved with constant speed in a circular path with the Earth at its center. Accordingly, in (a) the inner curve is a circle with the Earth E at the center. The inner tick marks divide the circle into twelve equal parts, representing the positions of the Sun at equal time intervals according to Aristotle's theory. It is a discrepancy with Aristotle's theory that, as viewed from point E, the outer tick marks, representing the *actual* positions of the Sun, do not line up with the inner tick marks, representing the *theoretical* positions of the Sun. This nonconstant speed of the Sun entails unequal lengths of the four seasons, a condition observed by Hipparchus. The current lengths in days of the seasons are spring: 92.764; summer: 93.647; autumn: 89.836; and winter: 88.897.

Hipparchus reduced the discrepancy of Aristotle's model, as shown in (b), by positing that the Sun's orbit is a circle centered, not at E, but rather at O, a point displaced from E in the direction of aphelion (roughly the constellation Gemini) at a distance equal to 1/24 of the radius of the circular orbit—an *eccentric* circular orbit. The inner and outer tick marks are now in closer alignment—an improvement in modeling the Sun's motion.

Figure 2.9b shows how Hipparchus reduced this discrepancy by making the inner circle slightly eccentric—that is, by displacing the center from point E to O. Note that now the inner and outer tick marks are more nearly lined up, showing closer agreement with the motion of the Sun.

Mean solar time

The rising and setting of the Sun is a metaphor for the inevitable regularity of time. The system of time based on the movement of the Sun is called *apparent solar time.* The sundial was the device that determined apparent solar time for daily activities before today's universal use of accurate clocks. However, in the previous section we have seen that the elliptical movement of the Sun is not entirely regular—it is fastest at perihelion (currently in early January) and slowest at aphelion. A second source of the irregularity of apparent solar time is the *obliquity of the ecliptic*, the angle between the planes of the equator and Earth's orbit about the Sun, which causes the apparent Sun to move faster near the summer and winter solstices and slower near the spring and autumn equinoxes.

In the seventeenth century, mechanical clocks of great accuracy were invented. On ocean voyages, latitude could be determined by examining the stars, but to determine longitude, accurate clocks were necessary. Accurate clocks showed clearly the irregularity of apparent solar time and led to the definition of *mean solar time*, which was based on a fictional Sun, the *mean Sun*. The mean Sun circles about the celestial equator at a constant rate while the true Sun circles about the ecliptic at an uneven rate. The mean Sun orbits the Earth's equator at 1 revolution per 365.2564…days, while the true Sun completes one revolution of the ecliptic in the same period of time at an uneven rate. To define mean solar time unambiguously, one can arbitrarily choose that at some particular moment mean solar time agrees with apparent solar time. By the current convention, equality occurs on December 25, which implies equality also on April 15, June 13, and September 1, dates of no particular astronomical significance.

Mean solar time, the system of timekeeping based on the mean Sun, is the familiar system of time that today runs the world's clocks. The difference between mean solar time and apparent solar time is called the *equation of time*.[10]

The sundial and the analemma

A sundial consists of an upright arm, the *gnomon*, that casts a shadow that points to the time of day on a horizontal dial calibrated suitably depending on the geographical latitude.

At night, the Sun is not visible, but one can use the fixed stars as a means of telling time. In fact, the movement of the stars, based on the rotation of the Earth, is a much better timekeeper then the sundial. Unfortunately, the brightness of the Sun makes it impossible to check the performance of the sundial compared to the stars. The rigorous checking of the timekeeping quirks of the sundial would have to wait for the invention of accurate mechanical clocks in the seventeenth century.

An *analemma* is a graphical device (Figure 2.10) that provides a correction to the sundial depending on the date. The analemma shows the departure of the apparent Sun from the mean Sun at each day of the year. The horizontal coordinate is the lead or lag of the apparent Sun over the mean Sun (the equation of time), and the vertical

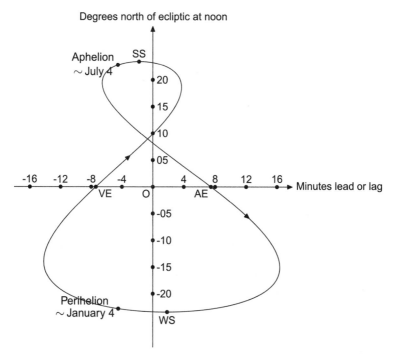

Degrees north of ecliptic at noon

Figure 2.10 The analemma, showing the departure of the true Sun from the mean Sun depending on the date. On the horizontal axis, minutes of lead or lag can be converted to degrees of right ascension (4 clock minutes = one degree of longitude). The vertical axis measures the declination of the true Sun. VE = vernal equinox, SS = summer solstice, AE = autumnal equinox, WS = winter solstice.

component shows the declination of the apparent Sun. A single analemma serves for sundials at all geographical latitudes.[11]

An analemma is frequently displayed on globes of the world as in Figures 2.11a and b. Note that the analemma is placed on the globe so that its scale in degrees coincides with the latitude and longitude of the globe, whereas in Figure 2.10, the horizonal scale, measuring the time in minutes, is stretched relative to the vertical. (One degree of angle is equal to $24 \times 60/360 = 4$ clock minutes.)

If the Sun is photographed using multiple exposures at exactly the same time every day, the resulting image traces the analemma as in Figures 2.11a and b. The photographic image is vertical, as in Figure 2.11a, only if the exposures of the Sun are taken exactly at noon. When the exposures are made at 4:00 p.m. instead of noon, Figure 2.11b shows that the resulting analemma is tilted with respect to the horizon. In Figures 2.11a and b, P denotes a point at latitude 40° North.[12]

One of the works of Ptolemy bears the title *Analemma*. At that time, "analemma" meant merely a graphical tool for calculation.

The analemma as a device for correcting the time generated by a sundial came into being only in the eighteenth century.[13] The analemma was not only computed

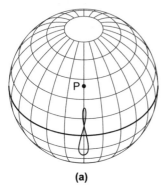
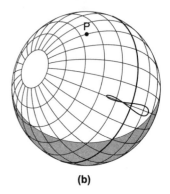

(a)   **(b)**

Figure 2.11 Figures (a) and (b) both represent the celestial sphere. Point P represents the zenith from a point on the Earth at 40° North (e.g., Philadelphia)—at noon in (a) and at 4:00 p.m. in (b). Multiple exposures of the Sun taken at the same time each day for a full year produces an analemma as an image. In (a), the Sun is photographed exactly at noon each day, and in (b) at 4:00 p.m. In (a), the entire view shown is above the horizon. In (b), the part of the celestial sphere below the horizon and not visible from P is shaded. The analemma in (b) appears slanted with respect to the horizon.

theoretically, but it was also observed experimentally through the use of accurate clocks.

This chapter has shown that the two-sphere geocentric model of the universe is adequate to describe the motion of the fixed stars and the Sun. In the next two chapters, we will see how the ancient Greeks devised geocentric methods for predicting the motions of the Sun, Moon, and planets that continued in use until the sixteenth century.

3

EPICYCLES AND RELATIVE MOTION

The ancient Greek astronomers used *epicyclic motion* to give a geocentric description of the motions of the planets. They proposed this type of motion for two reasons:

1. Epicyclic motion satisfied Plato's dictum that the motion of the planets should be based on uniform circular motions.
2. Epicyclic motion gave a remarkable good fit with the observed motions of the planets.

This chapter and the two following ones deal with epicyclic motion. Chapter 4 considers how the ancient Greeks made use of epicyclic motion in their astronomy, and chapter 5 considers some decorative uses of epicyclic curves. However, the present chapter treats epicyclic motion as a geometrical topic as it might be presented in a modern classroom.

The ancient Greek astronomers were unable to see that the concept of *relative motion* could be used to understand epicyclic motion. Relative motion, now considered an elementary concept, is used in this chapter to show how epicyclic motion can be transformed into circular orbital motion.

This chapter looks at epicyclic motion through modern eyes. In the next chapter, we will see how Ptolemy and others used observations to determine the epicyclic motion of the planets.

3.1. A MECHANICAL LINKAGE

Epicyclic motion can be generated mechanically by a linkage of hinged rods as shown in Figures 3.1 and 3.2. In Figure 3.1, the segments EO and OM are rods of fixed lengths r_2 and r_1, respectively, hinged at O. The linkage EOM lies in a two-dimensional plane. Point E is marked with a star to indicate that it is the fixed point of the linkage. A hinge at the point of attachment permits the rod EO to rotate about the fixed point E. When the linkage moves, the rate of rotation of each rod is defined as the rate of change of the angle between the rod and the horizontal direction. Point M executes an epicyclic motion when both rods rotate with uniform angular velocity.

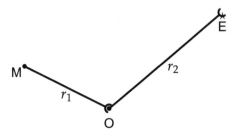

Figure 3.1 Two-rod linkage with fixed point of attachment at E and hinge at O. When the rods move with constant rates of rotation, the point M executes the epicyclic motion shown in Figure 3.2. Point E is marked with a ⋆ to indicate that it is the fixed point of the linkage. The point M holds a pen that draws the epicyclic curve $\mathcal{M}$, the path of the epicyclic motion.

Figure 3.2 shows the two-rod linkage of Figure 3.1 moving so that point M executes an epicyclic motion along the epicyclic curve $\mathcal{M}$. The linkage moves clockwise through ten positions spaced at equal time intervals. The hinge point O moves clockwise with constant clockwise angular velocity ω_2 along the circle $\mathcal{D}$, called the *deferent* circle. Point M moves clockwise with constant angular velocity ω_1 along the circle $\mathcal{E}$, called the epicycle, while, simultaneously, O, the center of circle $\mathcal{E}$, executes the circular motion described in the previous sentence. The motion of M, the composition of the two circular motions described above, is an epicyclic motion along the epicyclic curve $\mathcal{M}$.

The speed of the tracing point M is indicated by the closeness of the dots on the curve $\mathcal{M}$—slowing down on approaching the loop $\mathcal{L}$. The shape of the resulting curve $\mathcal{M}$ is determined by two ratios: the lengths of the rods $r_1 : r_2$ and the rotation rates $\omega_1 : \omega_2$.[1]

By adjusting the values of r_1, r_2, ω_1, and ω_2 one can obtain an arbitrary two-rod epicyclic motion. In general, an epicyclic motion can be defined with a linkage having an arbitrary number of rods, but in this chapter we only examine the case of two or three rods.

The Fundamental Theorem of epicyclic motion

I call the following theorem *fundamental* because it plays an important role in the development of planetary astronomy. It is a result that is easy to state and easy to prove. Section 6.1 mentions how Ptolemy made a passing reference to this result. However, Ptolemy did not see the deep importance of this theorem.

Surprisingly, the order of the rods does not effect the shape of the epicyclic curve. If the order of attachment of the rods is changed, the same epicyclic curve is generated, provided that each rod keeps its original rotation rate and its initial angle. More formally, we have the following theorem:

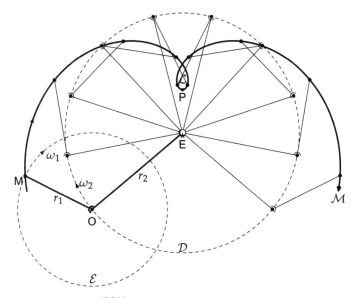

Figure 3.2 The two-rod linkage EOM, also shown in Figure 3.1, gives a clockwise epicyclic motion to point M along the epicyclic curve $\mathcal{M}$. The rods with lengths r_1 and r_2 rotate clockwise with angular velocities ω_1 and ω_2, respectively. The position of the linkage is shown going through ten equally spaced time intervals corresponding to ten equal arcs on the deferent circle $\mathcal{D}$ as point O moves with uniform clockwise angular velocity ω_2 along the deferent circle $\mathcal{D}$ of radius r_2. Point O is the moving center of circle $\mathcal{E}$, the epicycle, which rotates with uniform clockwise angular velocity ω_1. The point M on the circumference of the epicycle $\mathcal{E}$ executes the epicyclic motion $\mathcal{M}$.

Theorem 3.1 (The Fundamental Theorem of epicyclic motion). An epicyclic motion is unchanged if the order of attachment of the rods is changed, provided:

1. the lengths of the rods are unchanged,
2. the initial angle of each rod is unchanged, and
3. the rotation rate of each rod is unchanged.

This fact can be seen in the case of two rods by examining Figure 3.3, which shows that the two linkages, the original linkage EOM together with the reverse order linkage ESM, form a parallelogram. The vertex M of the parallelogram holds the pen for both linkages.

In Figure 3.3, the angle $\angle$SEO changes as time progresses, but the two linkages continue to form a parallelogram. In other words, EOM and ESM continue to meet at point M.

Thus, Figure 3.4 shows that the same epicyclic curve is obtained when the epicycle and the deferent and the corresponding rotation rates, ω_1 and ω_2, in Figure 3.2 are exchanged one for the other. Note that the point M travels around the deferent $\mathcal{D}$ so fast that more than one revolution of the deferent is required to draw the curve. To

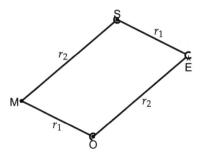

Figure 3.3 The two linkages EOM and ESM form a parallelogram.

avoid confusion, Figure 3.4 shows the construction of fewer points, only 6, whereas Figure 3.2 shows 10.

Orbital data for Earth and Mars are used in constructing Figures 3.2 and 3.4. In fact, these curves represent the motion of Mars according to the *deferent-epicycle model*, discussed in Section 4.3. As we will see later, in the deferent-epicycyle model, E is the Earth, M is Mars, and $\mathcal{M}$ is the path of Mars. Point O does not mark the position of any celestial body. Point S is of special interest. In Ptolemy's model, the Sun lay on the extension of the line segment ES in Figure 3.4. After Copernicus—for example, in the geocentric model of Tycho Brahe—the Sun was placed precisely at point S.

The argument used to establish Theorem 3.1 implies a more general result. The same path is traced if the deferent and epicycle are interchanged even if:

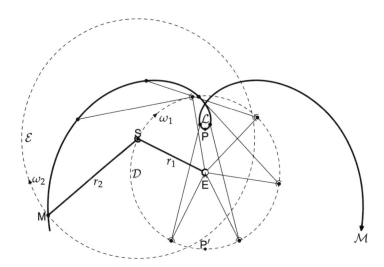

Figure 3.4 The epicyclic curve in Figure 3.2 drawn with exchange of deferent and epicycle. The fact that these two epicyclic curves are identical but differently generated illustrates Theorem 3.1. Orbital data for Earth and Mars are used in constructing these curves. In fact, these curves represent the motion of Mars according to the *deferent-epicycle model*, discussed in Section 4.3. The fact that these two epicyclic curves are identical but differently generated illustrates Theorem 3.1.

1. the deferent and epicycle are traversed at speeds that are not constant,
2. the deferent and epicycle are not circles,
3. and even if the deferent and epicycle are not two-dimensional curves.

For example, it is true that interchanging the order of the deferent and epicycle yields the same path even if the deferent and epicycle are planetary orbits as they are understood today—ellipses, not necessarily in the same plane, with speeds that are not constant.

The definition of epicyclic motion can be extended to include linkages with arbitrarily many rods. The rods are connected head-to-tail like vertebrae in a spine. For example, Figure 5.2 shows a three-rod linkage. The tail of the first radius is fixed. The rods rotate in a fixed plane, each with a constant rotation rate independent of the others. The head of the last radius, unattached to any other radius, holds a pen and draws the epicyclic curve.

3.2. RELATIVE MOTION

A *frame of reference* is a coordinate system, which is assumed to be stationary over time and which is used as a standard for measuring motion. In order to chart the motion of the planets, one must relate that motion to a frame of reference. Until the time of Einstein at the turn of the twentieth century, astronomers assumed that there was but one correct frame of reference—be it geocentric or heliocentric. However, AE (After Einstein) one is inclined to allow variety and to agree that well-defined frames of reference, however awkward, are logically admissible. It must be understood, however, that relative motion was an alien concept to Ptolemy and the other ancient astronomers. The ancient astronomers preferred a geocentric frame of reference in which Earth is the unmoving center—disagreeing with modern astronomers who have compelling reasons to give the Sun this role. There is no logical flaw regarding either Earth or the Sun as motionless and charting the motion of the rest of the solar system accordingly.

"Mars doesn't *really* travel along a loopy epicyclic curve like Figure 3.2, does it?" Looking at Mars in the night sky for an hour or more with the naked eye, one does not perceive any motion. However, over a period of a month it will be obvious that Mars has moved relative to the fixed stars. This motion is generally eastward with occasional westward retrograde episodes as shown in Figure 4.5. The answer to the question at the beginning of this paragraph is, Yes, as viewed from Earth, relative to the fixed stars, Mars does occasionally exhibit retrograde looping motion.

Observation from Earth of the position of a planet has just two dimensions—for example, azimuth and elevation, or right ascension and declination. Before current space technology, placing a planet in a three-dimensional coordinate system was only possible in some theoretical context—as provided, for example, by Ptolemy or Copernicus. Coordinate systems in current use are confirmed by actual space exploration.

A motion relative to a particular frame of reference can be described mathematically using an appropriate coordinate system. (See the discussion of coordinate systems in Section 2.2.) A different frame of reference results in a different mathematical description of the same motion. For example, Mars has epicyclic motion relative to a geocentric coordinate system and elliptical motion relative to a heliocentric coordinate system. There is more than one way to define these coordinate systems. The geocentric-equatorial and the heliocentric-ecliptic systems are discussed below. To specify a three-dimensional coordinate system, it is sufficient to specify the origin, the Z-axis, and the X-axis.

The geocentric-equatorial coordinate system uses the center of the Earth for its origin. The Z-axis is from the origin through the north pole. The X-axis is from the origin toward the vernal equinox—a point in the constellation Ares. Although the center of the Earth is fixed in this coordinate system, the surface of the Earth completes one rotation about its axis in one sidereal day (23.93 hours).

The heliocentric-ecliptic coordinate system uses the center of the Sun for its origin. The direction of the Z-axis is perpendicular to the ecliptic plane—the plane of Earth's orbit. The X-axis is directed toward the vernal equinox. In this coordinate system, Earth orbits the Sun once each sidereal year (365.2465 days).

The motion of the planets can be described fully in either of these frames of reference, but eventually it was discovered that the heliocentric frame of reference has the advantage of greater simplicity.

Heliocentric motion of Mars

The epicyclic motion of Mars in Figure 3.4 is greatly simplified if S, the Sun, instead of E, the Earth, is taken to be fixed—in other words, if the geocentric point of view is replaced by the heliocentric. The result of this change of viewpoint is shown in Figure 3.5. The linkage ASC is fixed at point S, the Sun, and the rods SA and SC trace the circles $\mathcal{E}$ and $\mathcal{D}$, the orbits of Earth and Mars, respectively. Note that the epicyclic path of Mars $\mathcal{M}$ in Figure 3.4 has disappeared entirely; Figure 3.5 contains no curves that are more complex than circles.

Figure 3.5 shows the positions of the linkage ASC at nine equally spaced instants of time. Thus, in the first time interval, point A (Earth) moves along the circular arc $\oversetfrown{AB}$ centered at S (Sun), and point C (Mars) moves along the arc $\oversetfrown{CD}$. This is a simplification of Copernicus's representation of the orbits of Earth ($\mathcal{E}$) and Mars

($\mathcal{M}$)—a simplification because Copernicus made small adjustments, to be discussed later, to improve the accuracy of this model.

In Figure 3.5, the arrows $\overrightarrow{AC}$, $\overrightarrow{BD}$, ... represent the line of sight from Earth to Mars at each of the nine equally spaced instants of time. This line of sight reverses its angular movement from clockwise to counterclockwise in the sixth and seventh time intervals—corresponding to the arc $\overarc{LMN}$—because this is where Earth overtakes Mars in its journey around the Sun. Earth passes Mars in this manner because, like the other superior planets Jupiter and Saturn, Mars moves on an orbit $\mathcal{D}$ larger than Earth's ($\mathcal{E}$) with an angular velocity with respect to the Sun that is less than Earth's. As Mars travels the arc $\overarc{LMN}$, Mars briefly reverses its movement in the sky relative to the fixed stars from eastward to westward. This *retrograde* motion causes loop $\mathcal{L}$ in Figures 3.2 and 3.4. Retrograde motion, discussed at length in the next chapter, exhibited by Mars and the other superior planets, was the major motivation for Ptolemy and other ancient astronomers to use epicycles as a model for planetary motion.

The ancient astronomers knew that that retrograde motion occurs only when a superior planet, Earth, and the Sun are close to alignment, as illustrated by the straight line SEM in Figure 3.5. This heliocentric figure shows that there is a geometrical reason for the connection between this alignment and retrograde motion. The ancient astronomers were unaware of this geometrical fact because it is not evident in geocentric models such as Figures 3.2 and 3.4—an important shortcoming of the epicyclic model. Figure 3.5, on the other hand, provides a geometrical explanation of the observed link between the Sun-Earth-Mars alignment SEM and retrograde motion.

In Figure 3.3, four candidates for the motionless center of the solar system present themselves. Earth and the Sun—that is, E and S in Figure 3.4—have already been

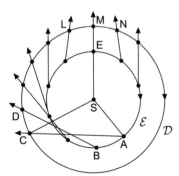

Figure 3.5 The motion of linkage ASC with point S, the Sun, held fixed. Points A, B, ... , E, ... and C, D, ... , L, M, N, ... represent the positions of Earth and Mars, respectively, at nine equally spaced moments of time—showing the clockwise movement of Earth and Mars around the circles $\mathcal{E}$ and $\mathcal{D}$, respectively. Figure 3.4 shows the same motion of Mars with Earth held fixed at point A in the above figure. The arrows represent the lines of sight from Earth to Mars at the nine equally spaced moments in time.

discussed. The other two choices are point M, Mars, and point O in Figure 3.2, which we will call the *empty center*. If Martian astronomers exist, they might choose Mars M to be the motionless center, but no terrestrial astronomers have advocated this choice.

The empty center

Holding the Sun motionless leads to modern astronomy as it has developed from the sixteenth century till now. However, it is an interesting exercise in the concept of relative motion if we examine a different choice. In Figure 3.2, point O plays a role similar to point S, the Sun, in Figure 3.4. What would happen if point O, the empty center, defined above, is held motionless? This would produce a device very similar to the orrery, described below. The difference is that point E, Earth, would rotate in a circle of radius r_2 with angular velocity ω_2, and Mars would rotate in a circle of radius r_1 with angular velocity ω_1—the motion shown in Figure 3.6 with points E and M, Earth and Mars, interchanged. Strangely, with respect to a motionless empty center O, Earth and Mars would exchange their radii, so that Mars would become an *inferior* planet with respect to the empty center O. The Sun S, on the other hand, would follow the epicyclic curve shown in Figures 3.2 and 3.4.

Ptolemy preferred Figure 3.2 over 3.4 as a description of the motion of Mars—a preference that puts the empty center in a prominent position. Conceivably, this could have misled Ptolemy or some later astronomer to look at planetary motion relative to a motionless empty center, as described in the previous paragraph. Such a scheme would have been more likely if Mars were the only visible planet. Since this is not so, each planet, apart from Earth, defines a different empty center. Thus, it would have been unlikely for the ancient astronomers to relate the motion of Earth to each of these empty centers.

Despite the indifference of pure logic, there are two advantages of the heliocentric system over the geocentric:

1. Charting the motion of the solar system with Earth motionless is much more complicated than the corresponding Sun-centered task.
2. The heliocentric solar system paves the way for Newton's theory of universal gravitation. The ancient Greeks studied the *kinematics*, the motion, of the solar system, but they had no corresponding theory of *dynamics*, of forces.

Today's astronomers prefer the heliocentric theory because the geocentric theory is more complicated without any compensating benefits. In fact, the most straightforward way to develop a correct geocentric theory of the solar system is first to develop the heliocentric theory and then to translate it back into the corresponding geocentric system. However, once the heliocentric theory is understood, it is a step backward to build a geocentric theory. Lacking this understanding, the infant science of astronomy took its first faltering steps.

The orrery—a mechanical model of the solar system

The rotating linkage describing the motion of Mars, shown in Figure 3.4, can be modified into a heliocentric linkage. This can be done by fixing point S, the Sun, instead of point E. When this is done, points E and M, Earth and Mars, rotate in circular orbits about S, the Sun, as shown in Figure 3.6.

With this modification, the linkage becomes an *orrery*, a mechanical device representing the motion of the planets, invented in 1710 by the English clockmaker George Graham and named in honor of his patron, the fourth Earl of Orrery.

An ordinary clock becomes an orrery if (1) the center of the clock represents the Sun, (2) the tip of the hour hand represents the motion of Jupiter, and (3) a point on the minute hand, suitably close to the center, represents Earth. This model works because the minute hand travels twelve times as fast as the hour hand, and the orbital period of Jupiter is 11.86 years, a number that happens to be close to twelve.

An orrery is inaccurate because it shows each planet moving with *constant speed* in a *circular orbit*. Nevertheless, this device is sometimes displayed in a science museum. Such a display is appropriate because, aside from its historical interest, an orrery gives a reasonable qualitative picture of planetary motion.

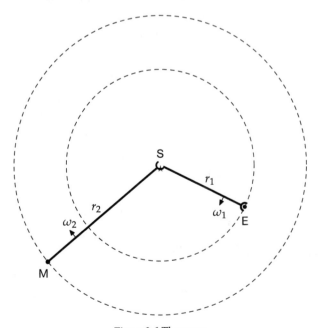

Figure 3.6 The orrery.

The deferent-epicycle model and the orrery describe the same motion—the former relative to a motionless Earth and the latter relative to a motionless Sun. Thus, if we say that the orrery gives a reasonable picture of planetary motion, then the same must be said of the deferent-epicycle model.

It is difficult for an orrery to incorporate the motion of all of the planets in the proper scale because, as shown in Figure 3.7, the orbits of the outer planets are so much larger than those of the inner planets.

The orrery is converted into a machine that illustrates the deferent-epicycle model if we break the connection that holds the representation of the Sun in place and nail down Earth. Physically, this might be difficult to accomplish because the motor that runs the orrery is generally attached to the Sun. Ignoring this technical problem, the motion that results is illustrated in Figure 3.8.

Since Saturn and Jupiter are so much farther from the Sun than Mars, Venus, and Mercury, Figure 3.8 breaks down the problem into two cases. Figure 3.8a shows the deferent-epicycle model applied to the motion of the outer planets (Saturn and Jupiter) and Figure 3.8b does the same for the inner planets (Mercury, Venus, and Mars).

The outer planets move at greater distances with slower speeds. For that reason the scale of Figure 3.8a is a six-fold magnification compared to Figure 3.8b. Moreover, the time period covered by Figure 3.8a is about six years (2,250 days), compared to about one year (375 days) for Figure 3.8b.

In Figures 3.8a and b, the path of the Sun is indicated by the dashed line. Linkages from Earth to the Sun and thence to each planet trace the movements of all of the planets. The rods all rotate clockwise with different uniform angular velocities.

Ptolemy would not have agreed with Figure 3.8 as a representation of the motion of the planets because he was misled by the Aristotelian notion of a set of crystalline spheres concentric with the Earth separating the motion of each planet from the others. Figure 3.8b would be unacceptable because the inner planets confusingly cross each other's path, whereas the crystalline spheres would keep them separate. Moreover, Figure 3.8a would be unacceptable because the crystalline spheres should ensure that there are no large gaps between the planets.

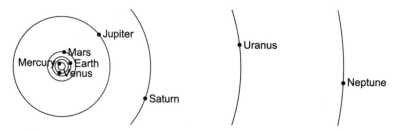

Figure 3.7 Orbits of the planets.

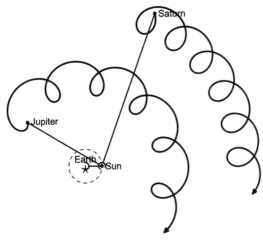

(a) Jupiter and Saturn over a period of about six years
(2,250 days).

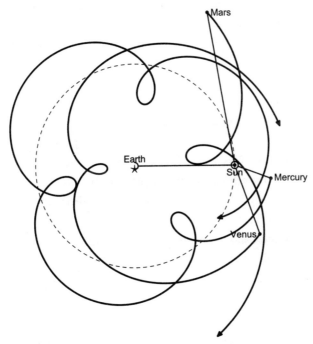

(b) Mercury, Venus, and Mars over a period of about one year
(375 days).

Figure 3.8 The geocentric motion of the planets according to the deferent-epicycle model.

The representations of Mars, Jupiter, and Saturn in Figure 3.8a ignore Ptolemy's preference that the deferent circle should be larger than the epicycle.

Figures 3.7 and 3.8 describe the same phenomenon, the motion of the solar system as a whole. The extreme complexity of Figure 3.8 compared to Figure 3.7 is yet another argument in favor of the heliocentric system compared to the geocentric.

Earthship 2

A child named Sue rides wants to ride the merry-go-round, but she is very sensitive to motion sickness. However, this particular merry-go-round has a seat in which she can enjoy the ride without suffering from vertigo because this seat gently rotates so that its occupant always faces north as the merry-go-round turns. As she rides, Sue decides to play a game, an exercise of the imagination, to suppress any residual motion sickness. She decides to imagine that she is not moving at all—that she is motionless, and that the entire merry-go-round is rotating about her. Joe—the operator and ticket taker—stands in the center, but, to Sue, he seems to be moving in a circle.

Sue holds a string attached to a toy airplane called Earthship 2. As she rides, she twirls Earthship 2 in circles over her head. Thus, it appears to Sue that she is motionless, and that Joe and Earthship 2 are each orbiting at different rates in circles about her, as shown in Figure 3.9a.

Sue, Joe, and Earthship 2 represent, respectively, the Sun, Jupiter, and Earth. Sue's game represents the heliocentric view of planetary orbits— apart from the fact that the true shape of the orbits is not quite circular (in fact, elliptical) and the speed of a planet in orbit is not quite constant.

Joe's view of the situation is different. He sees the path of Earthship 2 as compounded of two circles—the circular motions of the merry-go-round and Earthship 2. In fact, the path of Earthship 2 is a circle whose center is also moving on a circular path. Such a curve is a two-circle epicyclic curve. From Joe's (Jovian) point of view, Earth moves along an epicyclic curve. This two-circle epicyclic curve is generated by two circles: the circular path of Sue, the *deferent*, and the circular path of Earthship 2, the *epicycle*. Sue, at the center of the epicycle, moves along the circular path defined by the deferent. This is the view shown in Figure 3.9b.

Sue, a very inventive and imaginative young girl, soon tires of her first game and thinks of another. She supposes that Earthship 2 has a rider, her imaginary friend,

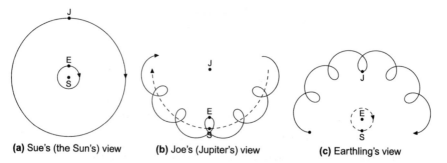

(a) Sue's (the Sun's) view   **(b)** Joe's (Jupiter's) view   **(c)** Earthling's view

Figure 3.9 Each of the three principals in the story of Earthship 2 can imagine that he or she is stationary and the other two are moving. The epicyclic curves were constructed from astronomical data for Jupiter. Figure (c) represents the geocentric form of deferent-epicycle model for the motion of Jupiter, and (a) is the corresponding heliocentric model.

Earthling. Because Earthling, like Sue, is highly susceptible to vertigo, she imagines that he also has a chair that always faces north. "What if Earthling plays my game? I mean, suppose that he thinks that he is at rest and I am moving in circles around him. Then how does he think that Joe is moving?" From his point of view, Sue is moving in a circular orbit, and Joe is moving in a circular orbit with Sue at its center. Earthling sees Sue follow a deferent orbit whose radius is equal to the length of the string she uses to twirl Earthship 2, and Joe traces an epicycle whose radius is equal to the distance between Joe and Sue. In other words, just as the ancient Greek astronomers believed that the path of Jupiter was an epicyclic curve, Earthling has these beliefs:

1. He (Earthling) is motionless;
2. Sue (the Sun) moves around him in a circular path;
3. Joe (Jupiter) follows the epicyclic curve shown in Figure 3.9c.

It is natural to ask, "If Earthling thinks that the path of Jupiter is an epicyclic curve, and the Jovians think that the path of Earth is also a curve of the same type, how do these two epicyclic curves compare?" These two epicyclic curves are *exactly the same!* More precisely, comparing Figures 3.9b and c, each curve on the celestial sphere is an antipodal reflection of the other. This is true because at each moment the line of sight from Earth to Jupiter is the reverse of the line of sight from Jupiter to Earth.

The story of Sue, Joe, and Earthship 2 is a simplified model for the relative motion of Jupiter, Sun, and Earth. Earthling's view, Figure 3.9c, is the deferent-epicycle model of the motion of Jupiter, and Sue's view, Figure 3.9a, is the corresponding heliocentric view. The two parts of Figure 3.8 are geocentric elaborations of Figure 3.9c.

The model represented by Sue's merry-go-round ride and by Figure 3.7 omits certain refinements by making the following simplifying assumptions:

1. The heliocentric orbits of the planets are concentric circles with the Sun at the center. The planets follow these circular orbits at uniform speed. The equivalent geocentric paths are two-circle epicyclic curves with Earth at the center of the deferent.[3]
2. The circular orbits (or the deferent and epicycle) lie in the ecliptic plane.

Early in the seventeenth century, Kepler knew that these two assumptions were only rough approximations. Kepler found that the orbit of Mars was a nearly circular ellipse with eccentricity about 0.09. (An ellipse with zero eccentricity is a circle.) His Second Law describes precisely how Mars moves slightly more slowly as it increases its distance from the Sun, and Kepler showed that the plane of the orbit of Mars has an inclination of about $1.9°$ with respect to the plane of the ecliptic.

To model planetary motion more accurately, the Earthship 2 merry-go-round needs to be modified to account for difficulties 1 and 2. Sue could easily tilt the orbit of Earthship 2 slightly. To make the orbits elliptical, the string holding Earthship 2 and the entire merry-go-round would have to be elastic. (They would have to stretch in a pre-scribed manner that also stretches the imagination.) Moreover, Sue's twirling motion and the merry-go-round itself would have to speed up and slow down appropriately.

These modifications of the story are far-fetched, but they are thinkable. Thus, the heliocentric or geocentric movement of the planets can be correctly modeled using minor modifications of the two-circle epicyclic curve. Thus, as we will see in chapter 6, Ptolemy and the ancient astronomers were tantalizingly close to a correct model of the solar system.

Ptolemy and the other ancient astronomers who supported the geocentric view of the solar system did not understand the relativism contained in the above story of Earthship 2. They did not have the flexibility to move back and forth between the geocentric and heliocentric points of view. If they could have done so, they would have realized that the heliocentric premise is simpler.

In this chapter, we looked with modern eyes at the tools of the ancient astronomers. In the next chapter, we will see how Ptolemy and the others determined suitable numerical values for the parameters of the epicyclic motion of the planets.

4

THE DEFERENT-EPICYCLE MODEL

The problem that ancient astronomers needed to solve was to form the correct theory of the *three*-dimensional motion of the planets from *two*-dimensional observations as in Figure 4.3. They succeeded rather well considering the difficulties—principally, the fact that the planets are so far away that simultaneous views from different observation points on Earth's surface are not observably different to the naked eye. At that time, parallax—discussed in Section 2.1—was not yet the useful astronomical tool that it was to become. At a later time, Kepler used parallax very cleverly (see Section 8.2) to track the motion of the planets.

4.1. RETROGRADE MOTION

Against the background of the fixed stars the motion of the planets is generally from west to east approximately following the ecliptic (defined in Section 2.3). However, it was troubling to Ptolemy and other ancient astronomers that occasionally a planet, when viewed from Earth, would exhibit *retrograde* motion—that is, would reverse its direction and move briefly westward. This section discusses an example involving the retrograde motion of Mars. The retrograde motion of Mars is easily observed because it happens when Mars is opposite to the Sun and at its brightest. On the contrary, the retrograde motion of Venus is difficult to observe because it only happens when Venus is close to the Sun in the sky. This distinction happens because Venus, together with Mercury, is an *inferior* planet, meaning that its orbit is closer to the Sun than Earth's. On the other hand, Mars, Jupiter, and Saturn are *superior*—having orbits farther from the Sun than Earth.

The *elongation* of a planet is defined as the angular distance, observed from Earth, between the Sun and the planet. The elongation of a planet changes with time. The inferior planets have limited elongation. The maximum elongations of Mercury and Venus are 28° and 47°, respectively, as shown, from the heliocentric viewpoint, in Figure 4.1. Consequently, these planets cannot be seen for more than three hours after sunset or before sunrise.

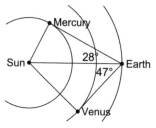

Figure 4.1 Maximum elongations of Mercury and Venus, respectively. The lines connecting Earth to Mercury and Venus are tangent to their respective orbits, which are approximated as circles.

The *inner* planets are, by definition, those closest to the Sun: Mercury, Venus, Earth, and Mars. The remaining planets Jupiter, Saturn—and, unknown in antiquity, Uranus, Neptune, and Pluto[1]—are called *outer* planets because they are much farther from the Sun. The inner planets are not the same as the inferior planets. Mars, in particular, is inner and superior. The inner planets are rocky, and the outer planets are gaseous—a fact not known to the ancient astronomers.

A retrograde motion of a planet occurs only when Earth, the Sun, and the planet are close to alignment. Using the modern heliocentric viewpoint, Figure 4.2 shows the terminology for alignments. Here, and elsewhere, we use the familiar heliocentric viewpoint when it serves to simplify the discussion—bearing in mind that Ptolemy and the other ancient astronomers, because they were committed to a geocentric universe, looked differently on these matters.

Note that the definitions in Figure 4.2 are different depending on whether the planet is inferior or superior. In all cases of alignment, the planet is situated such that its ecliptic longitude (defined in Section 2.2) is either the same as the Sun's or differs from the Sun's longitude by 180°. Since each planet's ecliptic latitude is small, this means that Earth, Sun, and the planet are nearly collinear at conjunction or opposition. For an inferior or superior planet, retrograde motion occurs when the planet is at *inferior conjunction* or at *opposition*, respectively. At inferior conjunction, inferior planets tend to get lost in the Sun's glare.

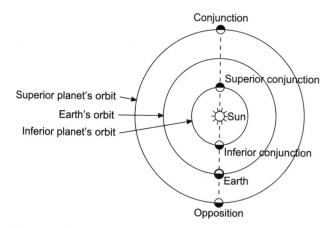

Figure 4.2 Alignments of inferior or superior planets with Sun and Earth.

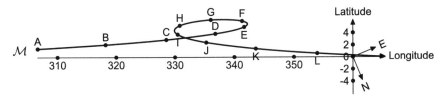

Figure 4.3 Curve $\mathcal{M}$ is a retrograde episode of Mars. The horizontal (longitude) axis coincides with the ecliptic circle. The vertical axis passes through the vernal equinox (longitude 0) at which point the eastward direction makes an angle of 22.5° with the ecliptic. The points A, B, ..., L on $\mathcal{M}$ represent sightings of Mars at 20-day intervals. The point G represents the position of Mars on August 27, 2003, an opposition during which that planet achieved its closest approach to Earth in almost 60,000 years. Note the use of the *celestial* compass, the mirror image of the terrestrial compass.

The orbit of each planet lies in a plane that makes a small angle, the angle of inclination, with the plane of the ecliptic. These angles are all about 2° or less except Venus (3.4°), Mercury (7.0°), and Pluto (17.1°). This is the reason why a planet need not coincide exactly with the Sun during a conjunction. If an inferior planet (Mercury or Venus) actually crosses the Sun, the inferior conjunction is called a *transit* of the Sun.

Figure 4.3 is an instance of retrograde motion that occurred during the weeks before and after the opposition of Mars that occurred on August 27, 2003. At that time, Mars achieved its closest approach to Earth in almost 60,000 years, although Mars has been very nearly as close many other times— most recently, on August 23, 1924. Mars appears to make similar, but not identical, loops at every opposition—events that occur about every twenty-two months.

Figure 4.3 shows the retrograde motion of Mars seen from the surface of Earth. Ptolemy and the other ancient astronomers were fairly successful in accounting for the movement in longitude of the planets—the horizontal movement in Figure 4.3. Ptolemy also accommodated latitudinal motion by using epicyclic curves with a small angle between the planes of the deferent and the epicycle—a technique that will not be discussed here.

4.2. PTOLEMY

The Alexandrian astronomer Claudius Ptolemy (85?–165? CE) wrote extensively on the motions of the Sun, Moon, and planets. Much of this writing represents the work of ancient astronomers who preceded him. Although he did not invent epicyclic curves, he applied them extensively.

Ptolemy titled his work *The Mathematical Compilation*, which today is known as *Almagest*, "The Greatest," an honorific given by Arab scholars who first translated this work from Greek to Arabic in the tenth century. In Europe, *Almagest* was translated into Latin from the Arabic in the twelfth century. It was held to be the highest authority on astronomy until the sixteenth century. *Almagest* is available in a recent English translation (Ptolemy, 1998). Ptolemy's other surviving works; include *Geography*, containing maps of the world known to Ptolemy; and *Tetrabiblos*, a work on astrology. Ptolemy's works *Analemma* and *Planisphaerium* deal with techniques of celestial measurement.

There are certain mysteries regarding Ptolemy. Nothing is known about his personal life. It has even been suggested that he lived much later because the star Polaris enjoys the first place in his star catalog although, when Ptolemy is said to have lived, Polaris did not mark the north celestial pole as it does today, and hence there was no reason to give that star a position of honor. However, there is no way to know whether the star catalog was "corrected" at a later date.

Ptolemy has been recently accused of fudging data and of plagiarism (R. R. Newton, 1977). However, this chapter is not dependent on the *source* of Ptolemy's astronomy.

Ptolemy and other ancient astronomers who used epicyclic curves to describe planetary orbits felt that they were following the advice of Plato who recommended that astronomers base their theories on uniform circular motion. The story of Earthship 2 in Section 3.2 illustrated how, relative to a fixed Earth, the orbits of the planets are approximately epicyclic curves.

One might think that because Ptolemy used the geocentric view, he would be unable to determine the motion of the planets based on a theoretical understanding of the solar system and that he must have used epicyclic curves merely as a *data-fitting* technique,[2] but this is not so, for the following two reasons:

1. Ptolemy used more than just the numerical data from his observations of the planets. He also connected the motion of the planets to the motion of the Sun. For example, he knew that at the midpoint of a planet's retrograde motion, the Sun, Earth, and the planet were in alignment.
2. A two-circle epicyclic curve was basically a correct model for a geocentric theory of planetary motion. The geocentric view is not incorrect—merely unduly complicated. This fact was illustrated in the story of Earthship 2 in Section 3.2.

One of the reasons for the persistence of the prejudice in favor of uniform circular motion was that Ptolemy succeeded too well in solving the problem posed by Plato. In fact, Ptolemy used epicyclic curves, based on uniform circular motion, to predict the motion of the planets with considerable accuracy. For more than a millennium, the fact that the errors were minor convinced Ptolemy's successors that he was on the right track and his methods merely needed refinement.

Ptolemy's mentor

It is not known who was Ptolemy's mentor, or, indeed, if he had a mentor. The following is a fictional account of what Ptolemy's mentor, if he existed, might have said to Ptolemy early in his career.

> Ptolemy, I have been impressed by your contributions to our seminar. In fact, I have the highest regard for your scientific abilities. At this time, I would like to suggest a research problem for you, provided that you are willing to undertake the risk of a truly difficult topic that might lead to a dead end.
>
> I have in mind the problem of *retrograde motion of the planets*. I'm sure that you have heard this problem discussed often in our seminar, but I feel that our current theories are inadequate.
>
> I have a chart here of a retrograde episode of Mars. (See Figure 4.3.) Here is a phenomenon that cries out for a comprehensive explanation. On the one hand, these retrograde loops occur with incredible regularity. On the other hand, the loops are never exactly the same, and they display remarkable asymmetry. If we get to the bottom of this, I feel sure that we will come to a fuller understanding, not only of the motion of the planets, but of every aspect of astronomy.

Ptolemy's mentor would be pleased if he could know that Ptolemy would become the most influential astronomer of the ancient world and that he would be considered the ultimate authority on astronomy until the sixteenth century. Ptolemy had the good fortune that his work *Almagest* was preserved and translated in the tenth century by Arab scholars. Ptolemy developed a great body of observational data and he demonstrated the predictive power of his methods. Ptolemy's work was based on careful study of detailed astronomical observations.

Ptolemy's mathematical models use epicycles, eccentrics, and equants. The theory of epicycles and eccentrics was known to Apollonius of Perga (262?–190? BCE) who lived more than three hundred years before Ptolemy. Ptolemy, himself, invented the equant, which became the basis of Kepler's *vicarious hypothesis* (Section 8.2)—the precursor of Kepler's epochal theory of elliptic orbits.

Ptolemy's greatest handicap was his insistence on the geocentric model of the solar system. The geocentric premise looks at the motion of the Sun, Moon, and planets relative to the motion of Earth. Was this incorrect or merely awkward? Ptolemy tried to use epicyclic curves to model the motion of planets, but he found that certain modifications were needed.

Ptolemy and his predecessors, Apollonius and Hipparchus, used epicyclic curves to model planetary motion. To achieve good agreement, the simplest version of this model—the two-circle epicyclic curve —needed modification. One of these modifications, the *eccentric*, is illustrated in Figure 2.9. Ptolemy's modification, the *equant*, will be discussed in connection with Figure 4.11.

4.3. THE DEFERENT-EPICYCLE MODEL

Figure 4.3 poses some difficult problems. We first consider an attempt, elegant in its simplicity but crude in its agreement with observation, to explain Figure 4.3.

It was difficult for the ancient Greeks to explain retrograde planetary motion while maintaining their preference for a geocentric universe and honoring Plato's dictum that the motions of celestial bodies must be based on uniform circular motion. For this purpose, Apollonius of Perga suggested epicyclic curves. Ironically, Apollonius also wrote extensively on conic sections, including the ellipse, which Kepler later found to be the true shape of planetary orbits. Epicyclic motion was later modified by Hipparchus and Ptolemy to give truer agreement with observed planetary motion, but we postpone consideration of these refinements. This section deals with the simplest form of epicyclic model, the *deferent-epicycle model*.

The deferent-epicycle model embodied Plato's requirement for uniform circular motion with beautiful clarity, but, unfortunately, it needed some tinkering to account acceptably for the motion of the planets. Despite these shortcomings, the deferent-epicycle model is considered in detail here for two reasons;

1. It illustrates the methodology of the ancient astronomers without undue complexity.
2. The deferent-epicycle model led forward to the heliocentric theory of Copernicus.

The deferent-epicycle model asserts that the motion of each planet is governed by a two-circle epicyclic curve with Earth located at the center of the deferent circle. The tracing point on the epicycle travels with uniform circular motion relative to its center, which travels with uniform circular motion about the deferent circle. Figures 3.2 and 3.4 show two versions of the deferent-epicycle model. These figures are discussed in Section 3.1. In both figures, Earth and Mars are points E and M, respectively; the curves $\mathcal{D}$ and $\mathcal{E}$ are the deferent and epicycle, respectively. In both figures, the curve $\mathcal{M}$ has the same shape and is traced at the same rate by point M. In other words, interchanging the deferent and epicycle results in the same motion.

The epicyclic curve in Figure 3.2 is denoted $\mathcal{M}$ because it represents the path of the planet Mars. To see the plausibility of this representation, suppose that the Earth E is in a slightly different plane from the epicyclic curve $\mathcal{M}$. Then the foreshortened view of $\mathcal{M}$ in the vicinity of point P as seen from point E might resemble the actual observations of Mars shown in Figure 4.3.

The deferent-epicyle model, illustrated in Figure 3.2, held important clues, most of which would be understood only much later by Copernicus and astronomers who followed him. As described later in this chapter, the ancient astronomers determined ω_1, ω_2 with great accuracy and the ratio $r_1 : r_2$ with less accuracy by observing the motion of Mars. Ptolemy and other ancient astronomers knew that ω_1 and ω_2 were the average rotation rates around the ecliptic of the Sun and Mars, respectively. However, they did not understand that $r_1 : r_2$ was the ratio of the average distances from the Sun of Earth and Mars—as posited much later by Copernicus in his heliocentric theory. Not giving this meaning to $r_1 : r_2$ and the corresponding ratios for the other planets was a crucial lack for ancient astronomy, leading to failure to understand the relative distances of the various planets—ignorance of the correct proportions of the solar system.

Ptolemy defined the deferent-epicycle model so that in Figure 3.2 the line OM is parallel to the line connecting Earth and Sun. Using Theorem 3.1, this is equivalent to the assertion that in Figure 3.4 the line ES passes through the Sun. Using the Copernican proportions, not known to Ptolemy, a much stronger statement can be made—point S *is* the Sun.

In Figure 3.2, the exchange of the deferent and epicycle, as permitted by Theorem 3.1, curiously loses track of the motion of the Sun S. In Figure 3.4, S is the Sun, but in Figure 3.2, point O does not represent a celestial body, suggesting that Figure 3.4 is the more meaningful representation of the motion of Mars.

Ptolemy and the ancient Greek astronomers chose the deferent always to be the larger circle. This choice is suitable for the inferior planets (Mercury and Venus) but not for the superior planets (Mars, Jupiter, and Saturn). Unfortunately, the representation in Figure 3.4 was not their preference for the representation of the motion of Mars. It would have been a better choice because (1) it shows clearly the position S of the Sun, (2) it can accommodate two or more planets simultaneously, and (3) it can lead from the geocentric to the simpler heliocentric premise. The ancient astronomers did not see this opportunity to link the deferent-epicycle model more tightly to physical reality.

Nevertheless, we will follow Ptolemy's preference, shown in Figure 4.4, in our later discussion of Mars.

Are epicyclic curves the wrong model for celestial motion? In the previous chapter we saw that epicyclic curves with three rods assume a great variety of shapes. In fact, the theory of *almost periodic functions*—due to the mathematician Harald Bohr (1887–1951), brother of the physicist Niels Bohr—implies that, with sufficiently many rods, epicyclic curves can approximate with arbitrary precision any planar motion in the wide class known as *almost periodic*, a class that includes the motion of the planets.[3] Thus, it might not seem surprising that the ancient astronomers could fit epicyclic curves to the observed motion of the planets.

As noted above, the modern theory of almost periodic functions, unknown to the ancient astronomers, guarantees that additional epicycles—they are usually small, and I will call them *epicyclets*—generally improve the accuracy of the deferent-epicycle approximation of planetary motion. However, this additional accuracy comes at the cost of a misleading confusion! The deferent and epicycle are circles with a physical meaning, albeit unknown to the ancient astronomers. In fact, they are approximations of two heliocentric orbits—of Earth and a particular planet under discussion. Additional epicyclets, although they may improve the fit of the deferent-epicycle model to the observed planetary motion, do *not* have a similar independent physical meaning. Thus, the ancient astronomers' use of epicyclic curves with three or more generating circles improved the accuracy of the geocentric description of planetary motion, but this tended to obscure the next giant step forward—the Copernican revolution. Thus, the use of more complex epicyclic curves leads to a confusion between physical reality and an ingenious approximation technique. For this reason, this chapter discusses in greater detail the more physically meaningful two-circle model, the deferent-epicycle model.

The deferent-epicycle needs some tweaking, to be discussed later, to agree well with observations. For example, referring to Figure 3.2, putting the position of Earth at the center E of the epicyclic curve creates a pair of difficulties.

Difficulty 1.   If we, the observers, are located in the same plane as the epicyclic curve $\mathcal{M}$, the movement of Mars would be linear and we would be unable to see any retrograde loops. We would see Mars move back and forth along the ecliptic circle. The only hint of a loop would be Mars occasionally reversing its eastward path and increasing its brightness. In other words, this model describes the movement of Mars in ecliptic longitude and ignores the problem of latitude. The model needs some modification to account for latitudinal movement. For example, put the deferent and epicycle in slightly different planes. The resulting modification is a nonplanar curve.

Difficulty 2.   Although the deferent-epicycle model predicts multiple retrograde loops *of equal size*, careful observation of the planets shows that this is not the case. One method of dealing with this shortcoming is to modify the deferent-epicycle model by displacing Earth from its position at the center E of the epicyclic curve. This has the effect that, although the loops remain of equal size, they appear to be different when viewed from Earth. In Figures 3.2 and 3.4, point E represents Earth, the point from which Mars is observed. The accuracy of this representation can be improved by moving the position of E without changing the shape of the curve $\mathcal{M}$. This can be achieved by adding a third rod that is stationary—that is, with ω_3 equal to zero. This kind of tweaking helps, but Ptolemy found that a remarkable improvement resulted by introducing the *equant*, defined in Section 4.5. The equant is a device that permits a nonuniform rotation rate for one of the rods.

A stand-in for Mars

The two-circle epicyclic curve $\mathcal{M}$ in Figure 4.4 is a simplified stand-in for the more complex motion of Mars. Figure 4.3, on the other hand, depicts an actual retrograde episode of Mars. The stand-in $\mathcal{M}$ follows the deferent-epicycle model and uses Ptolemy's predilection for epicycles that are smaller than their deferents. $\mathcal{M}$ is defined as follows. A deferent circle $\mathcal{D}$ of radius r_2 rotates at the constant clockwise rate ω_2 and carries an epicycle of radius r_1 that rotates at the constant clockwise rate ω_1.

Figure 4.4 was drawn using modern astronomical data that were not known to Ptolemy. However, we will see how Ptolemy might have observed enough data to draw an approximation. This figure shows the epicyclic motion $\mathcal{M}$ together with the deferent $\mathcal{D}$ and four positions of the epicycle—$\mathcal{E}_1$, $\mathcal{E}_2$, $\mathcal{E}_3$, and $\mathcal{E}_4$. Mars's stand-in $\mathcal{M}$ uses parameter values that are derived from currently known averages for the motion of Mars and Earth. The radius r_1 is equal to 1 AU,[4] and the radius r_2 is equal to 1.52 AU, the current value for the mean distance of Mars from the Sun. The rotation rates

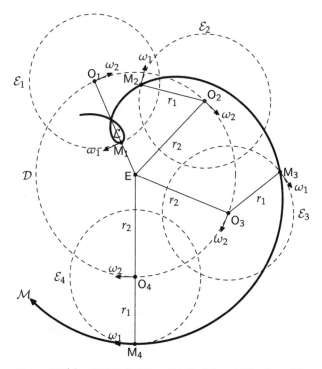

Figure 4.4 $\mathcal{M}$, a stand-in for Mars using the deferent-epicycle model.

are $\omega_1 = $ one revolution per year and $\omega_2 = 0.5337$ revolutions per year, the currently observed mean rotation rate of Mars about the Sun.

In summary, here are the parameter values used in the construction of $\mathcal{M}$:

$$\omega_1 = 1, \qquad \omega_2 = 0.5316 \text{ rev/yr} \tag{4.1}$$

$$r_2 : r_1 = 1.52 \tag{4.2}$$

Shortcomings and merits

This model does not explain a number of features of the actual observation of Mars shown in Figure 4.3. Although Figure 4.4 shows a retrograde loop $\mathcal{L}$, this loop—in fact, the entire path $\mathcal{M}$—can be seen only edge-on from point E because $\mathcal{M}$ is contained in the plane of the ecliptic. Thus, the view of the loop is one-dimensional. The retrograde episode is reduced to a brief westward excursion on a linear path that is generally eastward along the ecliptic—a gross simplification of the actual observation shown in Figure 4.3.

The remedy for the inability of this model to explain latitudinal motion is to limit the scope of the deferent-epicycle model to the explanation of longitudinal motion only.

Moving the point of observation slightly out of the plane of the curve $\mathcal{M}$ in Figure 4.4 would make the loop visible. However, it would be difficult to produce in this manner the asymmetry of Figure 4.3.

The actual retrograde loops of Mars occur in different sizes. Some are at least twice the size of others—a phenomenon observable with the naked eye. On the other hand, the deferent-epicycle model generates loops that are all the same size.

In view of these shortcomings, what are the merits of the deferent-epicycle model? Although the model does not describe latitudinal motion, it may still have merit in describing the longitudinal motion of the planets. From the Copernican heliocentric point of view, this model facilitates accurate computations of a number of important parameters of the planetary orbits—for example, the length of the Martian year and the relative distances of Mars and Earth from the Sun.

The ancient Greek astronomers were aware of the shortcomings of the deferent-epicycle model, and they introduced modifications to provide a better fit to the data. To this end, Hipparchus and Ptolemy used the *eccentric* epicycle (see Figure 3.9), and Ptolemy introduced the *equant* (discussed in Section 4.5).

Despite its defects, the deferent-epicycle model is worth discussing further because fitting the observational data to this model illustrates, in a simple setting, techniques that are used in more complex models. In the next two sections, observational methods are introduced that can be used to fit the deferent-epicycle model to planetary data. These methods are validated by showing that, applied to Mars's stand-in $\mathcal{M}$, they yield back the above numerical values of ω_1, ω_2, and the ratio $r_2 : r_1$ that were used in constructing $\mathcal{M}$. Stated more formally, in the next two sections we will solve:

Problem 4.1. Find terrestrial observations of $\mathcal{M}$ to determine ω_1, ω_2, and the ratio $r_2 : r_1$.

Determining ω_1 and ω_2

The deferent-epicycle model purports to explain retrograde motion. Therefore, it is natural to seek details concerning the recurrences of retrograde motion. The ancient Greek astronomers used observations of the times of *opposition* of the Sun and Mars.[5] These are midpoints of retrograde episodes of Mars, times at which the ecliptic longitudes of the Sun and Mars differ by $180°$. Figure 4.5 shows two successive oppositions of Mars according to the deferent-epicycle model. The time interval between two such oppositions is called a *synodic period*.[6] These events were charted in antiquity over many centuries. Ptolemy, for example, used both his own observations and records of old observations. Long before Ptolemy, the Babylonians knew that the motion of Mars is repeated, very nearly, in a 79-year cycle—that is, oppositions of Mars occur at

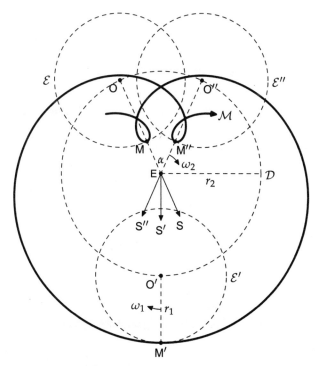

Figure 4.5 Finding the rotation rates of the Mars stand-in. Two successive oppositions, here are showing the deferent $\mathcal{D}$ and three positions, $\mathcal{E}$, $\mathcal{E}'$, and $\mathcal{E}''$ of the epicycle.

nearly the same longitude every 79 years. The synodic period of Mars is 779.94 days ($= 2.135$ years). The accuracy implied by these numbers is not unreasonable because the ancient astronomers used observations over centuries to get an accurate average. Mars's synodic period exceeds two years by 13.5%. This translates to a $48.6°$ increase in ecliptic longitude between successive oppositions.

Figure 4.5 depicts two successive oppositions of Mars at points M and M″ with Earth located at point E. The exact positions of the Sun are not shown in this figure, but, because points M and M″ are oppositions, the Sun must be in the directions MES and M″ES″ as indicated by the arrows. As noted above, a is equal to $48.6°$. It will be convenient to express a as a fraction of a revolution—that is, $a = 48.6/360 = 0.135$.

In Figure 4.5, point M′ represents the point of *conjunction* of Mars midway between the oppositions at M and M″. A conjunction is a point at which the Sun and Mars have the same ecliptic longitude. Observation of a conjunction is handicapped by the disappearance of Mars behind the glare of the Sun. The direction to the Sun is indicated by the arrow ES′.

The arrow ES, representing the direction of the Sun, rotates with uniform angular velocity about Earth (E) at the rate of one revolution per year. Thus, starting at ES the arrow rotates 2.135 revolutions clockwise to reach ES″.

Using Ptolemy's preference, we assume, as shown in Figure 4.5, that the radius r_1 of the epicycle $\mathcal{E}$ is smaller than the radius r_2 of the deferent $\mathcal{D}$. It follows that the length of EM is equal to $r_2 - r_1$.[7] Points O, O′, and O″ are positions of the center of the deferent circle. Arrows ES, ES′, and ES″ show the *directions* of the Sun from Earth at various times.[8] At the oppositions EM and EM″, Earth is between Mars and the Sun, and at the conjunction EM′, the Sun is between Earth and Mars. Thus, the rotation rate of the epicycle $\mathcal{E}$ is equal to the rotation rate of the Sun about the Earth—in other words, $\omega_1 = 1$ revolution per year.

The time between oppositions, $2 + a = 2.135$ years, is the time required for Mars to travel one revolution plus the angle a—in other words, 2.135 Earth years is equal to 1.135 Martian years. Noting that the interval from M to M′ contains just one retrograde loop—half at start and half at the end—this confirms the general relationship:

Retrograde loops + # Martian years = # Terrestrial years

This relationship is a special case of a more general formula. Suppose retrograde loops are observed looking from the surface of one planet to a second planet farther from the Sun—looking from an *inside* to an *outside* planet.[9] Then we have this equation:

Retrograde loops + # Outside planet years = # Inside planet years   (4.3)

Thus, according to the above calculation, a Martian year is equal to

$$\frac{2.135}{1.135} = 1.881$$

terrestrial years and, therefore, $\omega_2 = 1/1.881 = 0.5316$ revolutions per year.

In summary, we have solved half of Problem 4.1 by verifying the Martian rotation rates (4.1) of the deferent and epicycle, respectively.

Although the deferent-epicycle model is too crude for some purposes, we see here that, despite the handicap of its geocentricity, it is capable of determining the rotation rates of Mars and the other planets with a high standard of accuracy. The next task of Problem 4.1 is to determine the radii, r_1 and r_2. Actually, the naked eye observations available to the ancient astronomers are not sufficient to determine distances such as these. The underlying reason is that terrestrial distances are so small in comparison with astronomical distances that parallax is too small to be observed. However, the deferent-epicycle model and naked eye observations are sufficient to determine the ratio $r_2 : r_1$.

Determining the ratio $r_2 : r_1$

Figure 4.6 is a modification of Figure 4.4 that illustrates the method for finding the ratio $r_1 : r_2$. This method makes use of the fact that ω_1 and ω_2 are known.

Nightly observations of Mars permit the measurement of the longitude of Mars in its journey along the ecliptic, but such observation cannot confirm that Mars follows the particular path $\mathcal{M}$.

In Figure 4.6, M is a point of opposition of Mars. At that point, Mars M, Earth E, and the center O of the deferent circle lie on a straight line. Therefore, the angle EOM is $0°$. As time progresses, this angle increases because the rotation rate ($\omega_2 = 1$) of M about O is greater than the rotation rate ($\omega_1 = 0.5316$) of O about E. In fact, the rotation rate of angle EOM is equal to $\omega_2 - \omega_1 = 1 - 0.5316 = 0.4684$. It is possible to find a time at which the angle EO'M' is equal to $90°$, a quarter of a full revolution:

$$\frac{0.25}{\omega_2 - \omega_1} = \frac{0.25}{0.4684} = 0.5337 \text{ terrestrial years}$$

Measured in Martian years, this is equal to

$$0.5337\omega_1 = 0.5337 \times 0.5316 = 0.2837$$

Point O traces the entire deferent circle $\mathcal{D}$ in exactly one Martian year. Therefore, the angle OEO' is found by converting the time required for the journey of O to O'—found above to be 0.2837 Martian years—into degrees: $0.2837 \times 360 = 102.1°$.

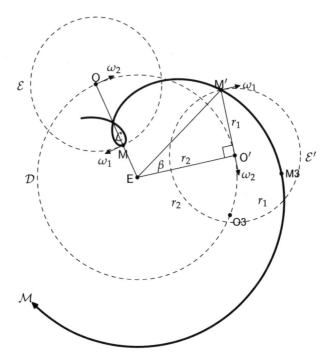

Figure 4.6 Determining the ratio $r_2 : r_1$ for the Mars stand-in.

Now we are ready to discuss the procedure for finding the ratio $r_2 : r_1$:

1. Observe the longitude a_1 of Mars at opposition.
2. Observe the longitude a_2 of Mars 0.5337 terrestrial years ($= 194.9$ days) later.
3. It is not possible to observe the longitude of O' because this point is a theoretical construct, not a celestial body. However, using the known values of ω_1 and ω_2, the above calculation shows that this longitude has increased $102.1°$ from its initial value at opposition.
4. Find the angle β in Figure 4.6 by subtracting the increase in longitude of Mars $(a_2 - a_1)$ from the increase in longitude of O' obtained in step 3.
5. Now the required ratio $r_2 : r_1$ is the ratio of the two legs of the right triangle $EO'M'$—in other words, the *cotangent* of the angle β.

For example, if β is observed to be $33.27°$, as in Figure 4.6, then this implies that the ratio $r_2 : r_1$ is equal to $\cot 33.27° = 1.524$. This solves the remaining part of Problem 4.1—verifying (4.2) using the observational methods of ancient astronomy.[10] Using actual measurements of the motion of Mars, good values can be obtained for the rotation rates. However, values of $r_2 : r_1$ are subject to greater error because the sizes of the retrograde loops vary considerably, depending on the ecliptic longitude at which they are observed. Good agreement with modern measurements of the mean distance of Mars from the Sun could only be obtained as an average over an entire cycle of 79 oppositions of Mars.

Ptolemy found estimates for ω_1, ω_2, and the ratio $r_2 : r_1$, but, because of his geocentric viewpoint, he did not interpret them as we do. Nevertheless, it can be said that the ancient astronomers had reasonable estimates for planetary rotation rates and the distances of planets from the Sun—even if they did not recognize them as such.

The other planets

The technique above for charting the motion of Mars was also used to study the motion of the other visible planets. Table 4.1 shows current values for the average orbital radii measured in astronomical units (AU) and the orbital period of the visible planets measured in sidereal years. Ptolemy calculated approximations of these values by the above methods, but he did not understand that these numbers were orbital radii and periods.

Ptolemy's choice that the deferents should always be larger than the epicycles produces unnecessarily complicated distinctions between the inferior and superior planets. For Ptolemy, the first column of Table 4.1 contains ratios between the radii of deferent circles and epicycles, and the second column contains the rates of rotation of deferents (in the case of the superior planets Mars, Jupiter, and Saturn) or epicycles (for the inferior planets Mercury and Venus). The epicycles of the superior planets, the deferents of the inferior planets, and the Sun all had the same rate of rotation: one year per revolution.

Triangulating the position of the Sun

It is a pervasive part, not just of modern science, but of art, literature, and other aspects of modern culture, that Earth is a planet like Mars and the other planets. We dream of visiting another planet, and soon this may be a reality. However, our present

Table 4.1 Current Data for Earth and the Visible Planets

Planet	Av. orbital radius in AU	Orbital period in sidereal years
Mercury	0.3871	0.2408
Venus	0.7233	0.6152
Earth	1.0000	1.0000
Mars	1.5237	1.8808
Jupiter	5.2034	11.8626
Saturn	9.5371	29.4475

fascination with space travel is fairly recent—for example, in *From the Earth to the Moon* (1865) by Jules Verne. The ancient world had accounts of fantastic travels but no planetary voyages. The idea that Earth is a planet was due to Copernicus and was unknown to Ptolemy and the other ancient astronomers. The following argument would have been a helpful step in the development of Copernicus's heliocentric theory, but it is unlikely that Copernicus used it.

The idea that Earth and Mars are both planets could lead to an argument that locates the position of the sun—by a kind of cooperation between Earthlings and Martians. This argument is needed only because we are adopting Ptolemy's preference—that the deferent should have a larger radius than the epicycle. Figure 4.7 shows Earth E, the deferent circle $\mathcal{D}$, and a position of the epicycle $\mathcal{E}$ that locates Mars at M. As noted above, the radius OM of the epicycle points in the direction of the Sun as seen from Earth. In other words, as seen from Earth, the Sun is in direction ES parallel to OM.

Now look at this situation from the point of view of the Martians. They also observe the Sun making a circuit of the ecliptic, a complete circuit in a Martian year. As noted above, the center O of the epicycle makes a circuit of the deferent $\mathcal{D}$ in a Martian year, 1.881 Earth years. The radius OE determines the direction of the Sun from Mars. In other words, the Martians see the Sun by looking in the parallel direction MS.

The two directions, from Earth and from Mars, intersect in S. In other words, S is the position of the Sun—found by triangulation, the intersection of the two directions.

The Sun makes a clockwise circuit once a year along the circle $\mathcal{E}'$. As seen in Figure 3.4, exchanging the epicycle and the deferent—putting the Sun S at the center of the epicycle—leads to the same path $\mathcal{M}$ for Mars.

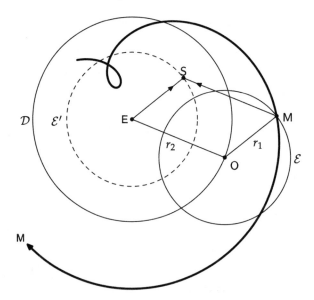

Figure 4.7 Finding the Sun.

Ptolemy and Copernicus

The ancient astronomers were handicapped by the clumsy geocentric viewpoint. The above line of thought was available to Ptolemy, but he and his contemporaries had no inkling that Earth was merely one of the many planets. Centuries after Ptolemy, Copernicus formulated a heliocentric solar system and gave reasonable approximations for the velocities and distances from the Sun for all the visible planets.

Ptolemy specified, incorrectly, a range of distances from Earth to the various planets—probably based on a useless speculation of Aristotle that a sequence of crystalline spheres existed centered about the Earth such that each consecutive pair of spheres was a "house" containing a particular planet. In that sense, there were not supposed to be any gaps between planets.

Heliocentric interpretation

The above geocentric methods for determining numerical values of ω_1, ω_2, and the ratio $r_2 : r_1$ have the simple heliocentric interpretations illustrated for the planet Mars in Figure 4.8. The rotational velocities ω_1 and ω_2 are the orbital velocities of Earth and Mars, and the ratio $r_2 : r_1$ is the ratio of the mean distances from the Sun of Mars and Earth—that is, the mean distance of Mars from the Sun in AU.

Figure 4.8b shows Mars in successive oppositions at ME and M″E″ with an intervening conjunction at M′E′. The angle α in Figure 4.8a is the same as angle α in Figure 4.5. The rotation rates ω_1 and ω_2 are computed as before.

Figure 4.8b shows Earth exceeding Mars by one quarter of an orbit. Angle β is the same as angle β in Figure 4.6, and $\gamma = 90° - \beta$. Figure 4.8b clarifies the fact that γ is the angle observed from Earth between Mars and the Sun. The required ratio is $r_2 : r_1 = \tan \gamma$.

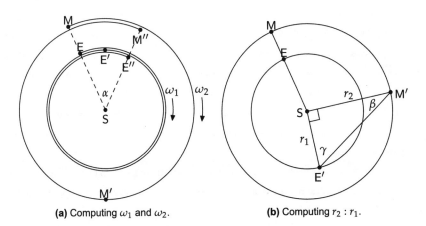

(a) Computing ω_1 and ω_2.   **(b)** Computing $r_2 : r_1$.

Figure 4.8 Heliocentric versions of Figures 4.5 and 4.6.

The comparison of Figure 4.8 with Figures 4.5 and 4.6 illustrates again the fact that heliocentricity clarifies the difficulties imposed by geocentricity.

4.4. INTIMATIONS OF NUMEROLOGY

Mars

It is an astronomical fact, known even to the Babylonians centuries before Ptolemy, that the close approaches of Mars to Earth, and the attendant retrograde loops, happen very nearly in 79-year cycles.

This behavior happens because the Martian year is equal to 1.88085 Earth years, a number that is close to $79/42 (\approx 1.88095)$. The number of loops is equal to the numerator minus the denominator of this fraction $(79 - 42 = 37)$.[11]

Venus

The recurrence of close approaches of Venus to Earth is even more striking. The radius—more properly, the semimajor axis—of Venus's almost circular elliptical orbit is 0.72333 AU, and the rotation period is 0.61529 years. This latter figure is closely approximated by $8/13 (\approx 0.61538)$. This approximation implies that Venus executes almost thirteen revolutions about the Sun every eight years during which time there are $13 - 8 = 5$ retrograde loops. Figure 4.9b shows the epicyclic curve of Venus continued for a period of eight years.

I think that Figures 4.9a and b are pleasing designs, and they may even seem expressive of the masculine and feminine, but there is no occult meaning here—only the fact that, by chance, certain decimals are approximated by certain fractions.[12]

As I write, we are passing through an unusual example of this cycle, the period between two transits of Venus, occurring on June 8, 2004, and June 6, 2012, but not again until December 13, 2117. A transit occurs when Venus crosses the Sun. During a transit, Venus is exactly between Earth and the Sun. In most of the eight-year cycles, Venus performs a near miss of the Sun, an inferior conjunction (see Figure 4.2) but not a transit. If the Venusians observe Earth at such a time, in the unlikely event of a clear day on that planet, they might say that Earth is in opposition, and they could observe Earth execute a retrograde loop. We could observe Venus do the same, were it not for the glare of the Sun.

The eight-year cycle of Venus is mentioned in Dan Brown's 2003 novel *The Da Vinci Code*. This cycle occurs in approximately thirteen Venusian years—as a consequence of equation (4.3):

\# Retrograde loops + \# Terrestrial years = \# Venusian years

(a)

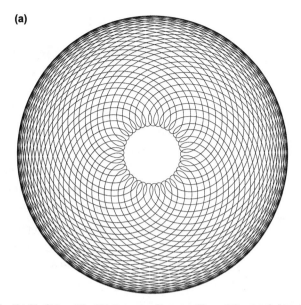

Figure 4.9a The Shield of Mars. The Martian epicyclic curve, Figure 3.2, extended to seventy-nine terrestrial years (about forty-two Martian years) with thirty-seven retrograde loops.

We obtain the mean rotation rate of Venus—or of Venus's deferent circle, as Ptolemy would say—as follows:

$$\frac{13}{8} = 1.625 \text{ revolutions per terrestrial year}$$

The true figure is 1.6254949 revolutions per year, an error of 0.03%.

(b)

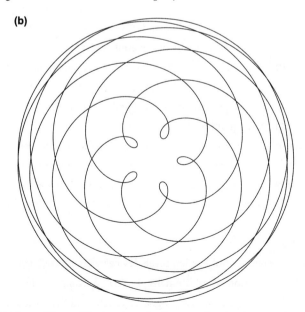

Figure 4.9b The Rose of Venus. The venusian epicyclic curve, showing the approximate eight-year cycle (thirteen orbits of Venus) with five retrograde loops.

4.5. THE EQUANT

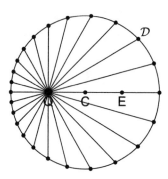

Figure 4.10 The equant point Q.

Ptolemy found it difficult to represent the motion of the Moon using epicyclic curves based on circles. Among the planets, Mars is a maverick, wandering off from the deferent-epicycle model more than most of the other planets. Ptolemy observed that a better fit sometimes resulted if the deferent circle was not traversed at a constant speed. He modified epicyclic motion, as shown in Figure 4.10, by requiring that motion around the deferent circle D have uniform rotation rate with respect to a point Q different from the center C of the circle, a point that he called the *equant*. Ptolemy's scheme required that Earth be located on the same diameter as Q and satisfy the condition $\overline{QC} = \overline{CE}$.

The resulting curve does not conform to our previous definition of epicyclic curves because it is not generated by a linkage in which each link executes a uniform rotation rate.

Copernicus and the earlier Arab astronomers found the equant a disagreeable complication—an annoying departure from the simplicity of uniform circular motion as dictated by Plato and Aristotle. Repairing this flaw was an important motivation for the Copernican heliocentric theory of the solar system.

We can thank Ptolemy for providing this prod to Copernicus, who was right to make the Sun the center of the solar system but wrong to think that he improved on Ptolemy's equant. As Kepler eventually discovered, it was better to abandon uniform circular motion. Ptolemy's equant was a remarkably close approximation of the correct elliptical orbit. Using an equant gives better accounting of the nonuniform speed of Mars in its orbit and the unequal size of the retrograde loops. Figure 4.11 is the epicyclic curve for Mars using the equant prescribed by Ptolemy.

The rejection of the equant by astronomers prior to Kepler carries an interesting lesson in scientific methodology. Science often uses concepts of beautiful simplicity, such as uniform circular motion, to explain nature. But nature always seems to be slightly more complicated than the philosophers and mathematicians would prefer.

The deferent-epicycle model of planetary movement was inaccurate and needed some kind of refinement. Until Kepler, three devices were offered: the eccentric, the equant, and the imposition of additional epicycles. A computer hacker might call them *kludges*—clumsy workarounds, but good enough until Kepler found the elegant solution.

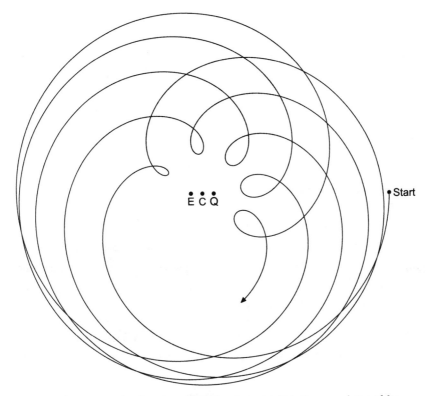

Figure 4.11 The Martian epicyclic curve using Ptolemy's equant. Note the unequal sizes of the retrograde loops.

Ptolemy's data for the Martian epicyclic curve (above)		
Radius of deferent	60.0	
Radius of epicycle	39.5	
Eccentricity	6.0	Distance from center C to equant point Q
		also distance from center C to Earth E
Deferent speed	0.52407116° per day	with respect to equant point Q
Epicycle speed	0.46157618° per day	with respect to center of moving circle

Refinements of the deferent-epicycle model improved agreement with astronomical observation. However, answers to the following two questions would have to wait for the Copernican heliocentric revolution.

1. Why are retrograde arcs of the superior planets centered on solar oppositions, as defined in Section 3.2?
2. Why do the inferior planets, Venus and Mercury, stay close to the Sun?

The next chapter discusses applications outside of astronomy, mainly decorative, of epicyclic curves.

5

MAKING MONEY, ET CETERA

...he made the universe a circle moving in a circle...
Plato, *Timaeus*

This chapter deals with the application and study of *epicyclic curves* outside of astronomy. We consider here how the ancient use of epicyclic curves foreshadowed curious applications to come—educational toys, designs to prevent counterfeiting of paper money, and challenging puzzles for seventeenth-century mathematicians.

- The *Spirograph*, an educational toy from Milton Bradley/Hasbro that was popular in the 1970s, is a device that enables the user, young or old, to design and draw intricate curves. Invented by Denys Fisher, a British engineer, the Spirograph uses geared wheels rotating on each other to draw epicyclic curves.
- Paper money, deeds, bonds, and stock certificates generally contain intricate patterns, called *guilloche*, to prevent counterfeiting. Guilloche patterns consist of epicyclic and related curves. The technique of guilloche was introduced by the Russian goldsmith and jeweler Peter Carl Fabergé (1846–1920) who used this method for decorating jeweled Easter eggs. He used a complex device called an *engine-turning machine* to engrave designs on metal objects. Today, engine-turning machines are rare, and guilloche patterns are created with the help of computer programs. Later in this chapter, we will see some other possibilities for epicyclic curves as decorative patterns.
- The use of epicyclic and related curves by mathematicians of the seventeenth century demonstrated the power of a newly discovered mathematical tool—the calculus of Newton and Leibniz. We discuss a few of their favorites: the cycloid, astroid, and cardioid.

The following section contains basic concepts that are central to later chapters.

5.1. EPICYCLIC CURVES AS DECORATIVE PATTERNS

Epicyclic curves are a rich source of patterns, many of them intricate. This section explores systematically the decorative use of epicyclic curves. The Spirograph is capable

Figure 5.1 An epicyclic curve as a guilloche pattern.

of drawing many epicyclic curves, but the computer graphics programming language Metapost, used for all the mathematical diagrams in this book, is a more sophisticated tool, capable of drawing epicyclic curves and much more.

The reader will find examples of similar decorative patterns within easy reach. Paper money makes use of patterns called *guilloche*, like the epicyclic curve in Figure 5.1. We will see how epicyclic curves can be constructed with sufficient symmetry and complexity to be used as guilloche. The principal tool for this analysis is a means of classifying epicyclic curves. Generally, guilloche patterns are not strictly epicyclic curves, although they are closely related, because they might be used to frame a noncircular region such as a rectangle.

Classifying epicyclic curves

Because three-circle epicyclic curves—for example, the guilloche pattern shown in Figure 5.1—are sufficiently complex for most decorative purposes, we begin with the classification of these curves. Referring to Figure 5.2, an epicyclic curve generated by a linkage OPQR of three rods attached at O is determined by the following parameters.

1. The three lengths of the rods OP, PQ, and QR are denoted r_1, r_2, and r_3, respectively. No unit of length need be specified if the scale of the figure is not important because only the proportion $r_1 : r_2 : r_3$ affects the shape of the curve.

2. The three rotation rates are denoted ω_1, ω_2, and ω_3, respectively. The choice of units (revolutions per year, degrees per second) only determines the time needed to trace the curve and has no effect on the shape of the curve. No unit of time need be specified if the time required to draw the curve is unimportant, as in a decorative pattern. Only the proportion $\omega_1 : \omega_2 : \omega_3$ affects the shape of the curve.

3. The three initial angles of the rods are denoted θ_1, θ_2, and θ_3, respectively. These angles are measured in degrees. With only two rods, the initial angles do not affect the shape of the curve as, in general, they do with three or more rods.

Thus the pattern generated by a three-rod linkage is determined by the nine numbers

$$(r_1, \omega_1, \theta_1; r_2, \omega_2, \theta_2; r_3, \omega_3, \theta_3)$$

which I will call a *signature* of the pattern. A signature defines a unique epicyclic curve; however, as remarked in items 1 and 2 above, it is possible that an epicyclic curve can be described by more than one signature.

The signature of a two-rod linkage consists of just six numbers, omitting r_3, ω_3, and θ_3. For example, a signature of the epicyclic curve in Figure 3.1 is

$$(1.52366, -0.53175, 90; 1, -1, -90)$$

The negative signs indicate clockwise rotation. Although the shape of a curve is unchanged whether it is traced in the clockwise or counterclockwise direction, the epicyclic curve in Figure 3.1 is traced in the clockwise direction to conform with the map-making convention that the direction west to east is shown left to right—in accordance with the usual (nonretrograde) movement of the planets with respect to the fixed stars.

In the examples to follow, we put $\theta_1 = 0$ and $\theta_2 = 0$. These conditions can be achieved by changing, if necessary, the time scale and/or rotating the figure about its center.

Table 5.1 gives the signatures of some of the epicyclic curves shown in these pages. Note that items (a), (b), and (g) represent exactly the same shape. In fact, items (a) and (b) are the same because the only difference is that the order of the two rods generating the epicyclic curve is reversed, and, as noted in Section 3.1, the epicyclic curve is not changed by changing the order of the rods. Further-more, the signature (g) (the Martian "shield") is different from (b) in two respects, neither

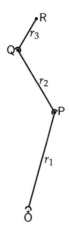

Figure 5.2 Three-radius linkage.

Table 5.1 Signatures of epicyclic curves

(a)	Figure 3.2	Martian epicyclic curve	$(1.52366, -0.53175, 90; 1, -1, -90)$
(b)	Figure 3.4	Martian epicyclic curve	$(1, -1, -90; 1.52366, -0.53175, 90)$
(c)	Figure 5.1	Guilloche	$(1, 10, 0; 0.094, 47, 0; 0.407, -64, 0)$
(d)	Figure 5.6b	Epicycloid	$(5, 1, 0; 2, 5/2, 0)$
(e)	Figure 5.7a	Astroid	$(3, 1, 0; 1, -3, 0)$
(f)	Figure 5.7b	Cardioid	$(2, 1, 0; 1, 2, 0)$
(g)	Figure 4.9a	Martian "shield"	$(1, 1, 0; 1.52366, 0.53175, 0)$
(h)	Figure 4.9b	Venusian "rose"	$(1, 1, 0; 0.72333, 1.62549, 0)$

of which affects the shape of the epicyclic curve: (1) The signs of ω_1 and ω_2 are plus instead of minus, and (2) the initial angles θ_1 and θ_2 are changed, producing, at most, a rotation. Figure 4.9a (the Martian "shield") has a different appearance from Figures 3.2 and 3.4 (the Martian epicyclic curve) only because the extent of the curve shown in the drawing is much greater for the former than for the latter. The signature represents the entire "infinite" extent of the epicyclic curve whereas drawings can only show a finite part—unless the epicyclic curve forms a closed curve, as discussed in the next section.

Closed epicyclic curves

It is natural to require that a decorative pattern should be a closed curve like Figure 5.6b (Table 5.1[d])—a condition satisfied if each of the rods executes a whole number of revolutions. Although it is difficult for the unaided eye to see that the guilloche pattern in Figure 5.1 (Table 5.1[c]) is a single closed curve, nevertheless it is also a closed epicyclic curve.

In Figure 5.6b the first radius executes two revolutions as the second radius makes $2 \times 5/2 = 5$ revolutions. Thus the signature $(5, 1, 0; 2, 5/2, 0)$ in Table 5.1(d) could be replaced by $(5, 2, 0; 2, 5, 0)$. Figure 5.6b—indeed, every closed epicyclic curve—has a signature in which the numbers ω_i are integers with no common factor. I will call a signature of this type *normal*. In the further discussion of closed epicyclic curves as decorative patterns, all signatures are assumed to be normal.

A normal signature is convenient because it exhibits the fact that the corresponding epicyclic curve is closed. Furthermore, if the signature is normal, then the entire curve is generated when the first radius executes ω_1 revolutions about the deferent circle—further movement about the deferent circle merely retraces the curve.

Two-circle epicyclic curves

Figure 5.3 shows a few closed two-circle epicyclic curves together with their normal signatures.

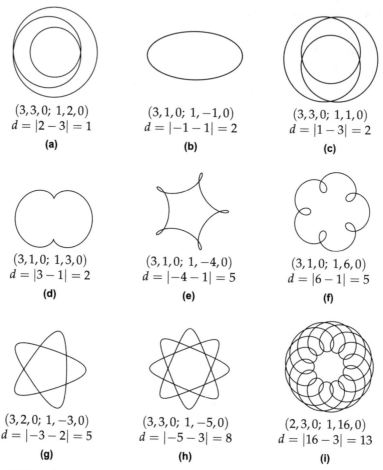

Figure 5.3 Some closed two-circle epicyclic curves; d denotes the multiplicity of rotational symmetry according to Proposition 5.1.

Pattern (b) is an ellipse. Any ellipse centered at the origin can be realized as an epicyclic curve. Kepler discovered that planetary orbits *are* ellipses, but with the Sun at a focus, not at the center. Furthermore, planetary orbits are traversed with a nonuniform speed not achievable with a two-circle epicyclic curve like (b).

The pattern (f), with inside loops, is similar to epicyclic curves that Ptolemy used to explain planetary motion, but pattern (e) with outside loops is not. Ptolemy preferred inside loops because they put Mars, for example, closer to Earth during retrograde motion—thus, explaining why Mars appeared brighter at those times.

Rotational symmetry of two-circle epicyclic curves

Closed two-circle epicyclic curves generally exhibit rotational symmetry. To say that a pattern has d-fold rotational symmetry, where d is a positive integer, means that the pattern will come into coincidence with itself when it is rotated about its center by the angle $360/d$ degrees. The central symmetry of two-circle epicyclic curves is characterized as follows:

Proposition 5.1. Let $(r_1, \omega_1, \theta_1; r_2, \omega_2, \theta_2)$ be a normal signature for a closed two-circle epicyclic curve. Put $d = |\omega_2 - \omega_1|$. If $d = 0$, then the epicyclic curve is a circle, a degenerate case. If $d = 1$, then the epicyclic curve has no rotational symmetry. Otherwise, the epicyclic curve exhibits d-fold rotational symmetry.

For example, Figure 5.3(h) has eight-fold rotational symmetry because

$$|\omega_2 - \omega_1| = |-5 - 3| = 8$$

All two-circle closed epicyclic curves display symmetry with respect to reflection across a line through the origin. In particular, the patterns of Figure 5.3 are symmetric with respect to the *horizontal* line through the origin. This is a consequence of the fact that, for these patterns, θ_1 and θ_2 are both equal to zero. Figure 5.3a exhibits symmetry by reflection across the horizontal axis, but no rotational symmetry since d is equal to 1.

The epicyclic curves shown in Figures 4.9a and b, representing the motions of Mars and Venus, respectively, are not closed. As previously noted, the shape of an epicyclic curve is determined by the ratio $\omega_1 : \omega_2$. Multiplying the ωs in the signatures (Table 5.1[c] and [d] by 79 and 8, respectively, we obtain the signatures

$(1, 79, 0; 1.52366, 42.0083, 0)$ and $(1, 8, 0; 0.72333, 13.0033, 0)$

for Mars and Venus, respectively. Thus, the epicyclic curves of Mars and Venus are closely approximated by the *closed* epicyclic curves with signatures

$(1, 79, 0; 1.52366, 42, 0)$ and $(1, 8, 0; 0.72333, 13, 0)$

These epicyclic curves have d-fold rotational symmetry with $d = 79 - 42 = 37$ and $d = 11 - 8 = 5$, respectively. This confirms the remarks concerning Mars and Venus in Section 4.4.

Three-circle epicyclic curves

The question of symmetry of three-circle epicyclic curves is more complicated. Figure 5.4 shows a selection of three-circle epicyclic curves with their normal signatures. I think the symmetric patterns are more decorative, and I have favored them in the selection of patterns in Figure 5.4. The three-circle epicyclic curves do not necessarily exhibit the reflective symmetry characteristic of the two-circle epicyclic

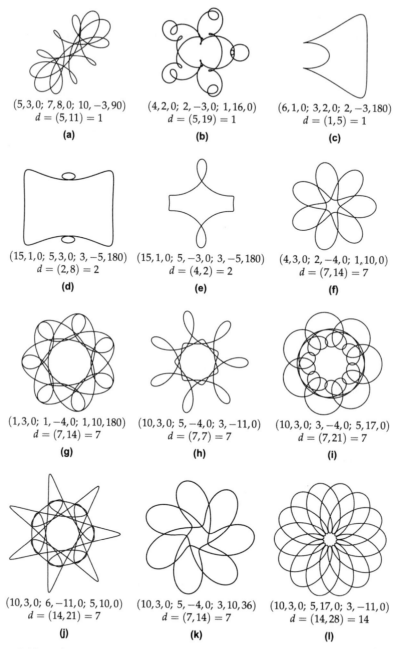

$(5,3,0; 7,8,0; 10,-3,90)$
$d = (5,11) = 1$
(a)

$(4,2,0; 2,-3,0; 1,16,0)$
$d = (5,19) = 1$
(b)

$(6,1,0; 3,2,0; 2,-3,180)$
$d = (1,5) = 1$
(c)

$(15,1,0; 5,3,0; 3,-5,180)$
$d = (2,8) = 2$
(d)

$(15,1,0; 5,-3,0; 3,-5,180)$
$d = (4,2) = 2$
(e)

$(4,3,0; 2,-4,0; 1,10,0)$
$d = (7,14) = 7$
(f)

$(1,3,0; 1,-4,0; 1,10,180)$
$d = (7,14) = 7$
(g)

$(10,3,0; 5,-4,0; 3,-11,0)$
$d = (7,7) = 7$
(h)

$(10,3,0; 3,-4,0; 5,17,0)$
$d = (7,21) = 7$
(i)

$(10,3,0; 6,-11,0; 5,10,0)$
$d = (14,21) = 7$
(j)

$(10,3,0; 5,-4,0; 3,10,36)$
$d = (7,14) = 7$
(k)

$(10,3,0; 5,17,0; 3,-11,0)$
$d = (14,28) = 14$
(l)

Figure 5.4 Some three-circle epicyclic curves; d denotes the multiplicity of rotational symmetry according to Proposition 5.2.

curves. For example, patterns (a) and (k) do not have reflective symmetry. I will not pursue a complete analysis of reflective symmetry for three-circle epicyclic curves. Suffice it to say that if the initial angles θ_1, θ_2, and θ_3 are each equal to either $0°$ or $180°$, then the epicyclic curve has symmetry with respect to the horizontal axis.

Rotational symmetry of three-circle epicyclic curves

If a pattern has the property that ω_1, ω_2, and ω_3 are terms of an arithmetic progression, not necessarily consecutive, with difference d, then the pattern has d-fold rotational symmetry. To assist in making the calculation of d, make the following definition where i and j are integers such that $1 \leq i < j \leq 3$:

$$d_{ij} = |\omega_i - \omega_j|$$

The greatest common divisor of positive integers m and n is denoted (m, n).

Proposition 5.2. *Let* $(r_1, \omega_1, \theta_1;\ r_2, \omega_2, \theta_2;\ r_3, \omega_3, \theta_3)$ *be a normal signature for a closed three-circle epicyclic curve, and suppose that* ω_1, ω_2, *and* ω_3 *are distinct integers. Let* $d = (d_{12}, d_{23})$. *If* $d = 1$, *then the epicyclic curve has no rotational symmetry. Otherwise, the epicyclic curve exhibits* d-*fold rotational symmetry.*

5.2. CYCLOIDAL CURVES

A special type of epicyclic curve is generated by a circle rolling inside or outside of another circle. The curve is drawn by a pen attached to the circumference of the rolling circle. If the circle rolls on a straight line instead of a circle, the resulting curve, shown in Figure 5.5, is called a *cycloid*. It is not a special kind of epicyclic curve. Together, these two types of curves are called *cycloidal* curves.

Nicholas of Cusa (1401–1464) was the first to study the cycloid. It was also studied by Marin Mersenne (1588–1648), and it was given its name by Galileo in 1599. Galileo tried, unsuccessfully, to find the area under a cycloid—even resorting to weighing cutouts. The list of those who wrote about the cycloid (Table 5.2) is a *Who's Who* of seventeenth-century thinkers. Quarrels ensued among them (e. g., concerning priority of various discoveries) to the extent that the cycloid has been called "the Helen of geometers," a reference to Helen of Troy, "the face that launched a thousand ships."

The mathematical pioneers of the seventeenth century were interested in the cycloid and the other curves discussed in this section, because they provided a testing ground for the blossoming ideas of the calculus of Newton and Leibniz.

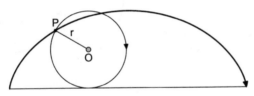

Figure 5.5 Cycloid.

Table 5.2 Seventeenth-century mathematicians who wrote about the cycloid.

Jacob (Jacques) Bernoulli (1654–1705)	Sir Isaac Newton (1643–1727)
Johann Bernoulli (1667–1748)	François Nicole (1683-1758)
Roger Cotes (1682–1716)	Blaise Pascal (1623–1662)
Girard Desargues (1591–1661)	Michelangelo Ricci (1619–1682)
René Descartes (1596–1650)	Gilles Personne de Roberval (1602–1675)
Pierre de Fermat (1601–1665)	René François Walter de Sluze (1622–1685)
Christiaan Huygens (1629–1695)	Evangelista Torricelli (1608–1647)
Phillippe de Lahire (1640–1718)	Vincenzo Viviani (1662–1703)
Gottfried Wilhelm von Leibniz (1646–1716)	John Wallis (1616–1703)
Guillaume François Antoine	Sir Christopher Wren (1632–1723)
Marquis de L'Hôpital (1661–1704)	

Cycloidal curves were studied by Ole Roemer (1644–1710), the Danish astronomer who first measured the speed of light. Roemer examined cycloidal curves in researching the best design for gear teeth: this was useful in countless mechanical devices.

A cycloidal curve is called a *hypocycloid* or an *epicycloid* depending on whether the rolling circle is contained in the fixed circle or not. The process of generating an epicycloid—a hypocycloid is similar—is illustrated in Figure 5.6. Figure 5.6a shows the construction of the epicycloid $\mathcal{E}$ with fixed and rolling circles $\mathcal{A}$ with radius a and $\mathcal{B}$ with radius b, respectively. In this example, the rods of $\mathcal{A}$ and $\mathcal{B}$ satisfy the proportion $a/b = 3/2$.

Figure 5.6b shows the construction of $\mathcal{E}$ using a linkage—demonstrating that the epicycloid $\mathcal{E}$ is an epicyclic curve with the deferent circle $\mathcal{D}$ of radius $a + b$. The motion

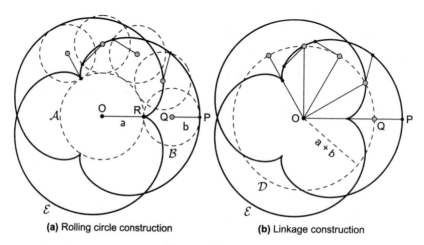

(a) Rolling circle construction   (b) Linkage construction

Figure 5.6 Epicycloid.

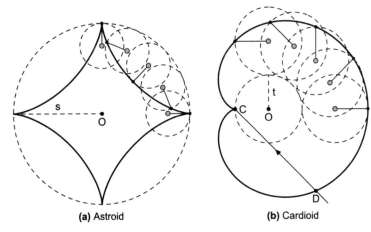

(a) Astroid   (b) Cardioid

Figure 5.7

of the linkage is determined by the condition that circle B rolls on circle A without slipping

A *hypocycloid* is generated similarly by a small circle rolling without slipping inside a larger circle. If the radius of the large circle is four times the radius of the small circle, the resulting hypocycloid, called an *astroid*, is shown in Figure 5.7a.

Finally, the heart-shaped epicycloid in Figure 5.7b is called a *cardioid*. A cardioid is generated when the stationary circle and the superior rolling circle are of the same size.

Cardioid microphones or antennas are designed with different sensitivities in various directions. Figure 5.7b represents a horizontal plane with the device placed at the cusp C of the cardioid. The line DC represents the path of a signal received by the device at C, and the length of the line segment DC is proportional to the amplification of that signal by the device. The sensitivity drops off to zero directly in back of the microphone or antenna. A directional cardioid microphone would not be good for a roundtable discussion, but it benefits a single speaker by increasing sensitivity on the speaker's side of the microphone, the right side in Figure 5.7b, while reducing any noise emanating from the opposite side, the left side.

Among the many remarkable properties of the cycloid, astroid, and cardioid, it can be shown, using calculus, that the arc lengths are integer multiples of the radii of the circles and the areas are simple multiples of the area of the fixed circle:

Cycloid (Figure 5.5)

Length of one arch $= 8r$ (4 times the diameter of the circle)
Area under one arch $= 3\pi r^2$ (3 times the area of the circle) cycloid

Astroid (Figure 5.7a)

> **Circumference** $= 6s$ (3 times the diameter of the fixed circle)
> **Area** $= \frac{3}{8}\pi s^2$ (3/8 of the area of the fixed circle) astroid

Cardioid (Figure 5.7b)

> **Circumference** $= 16t$ (eight times the diameter of the fixed circle)
> **Area** $= 6\pi t^2$ (six times the area of the fixed circle)

In this chapter, we defined epicyclic curves, and we found them to be a source of decorative patterns and of problems for the infant calculus. In the next chapter, we will see that these curves were the basis for Ptolemy's system for representing the motion of the planets—a system that dominated astronomy for more than a millennium.

PART TWO

Rebirth

6

THE RELUCTANT REVOLUTIONARY

In no other way do we find a wonderful commensurability and a sure harmonious
connection between the size of the orbit and the planet's period.
Copernicus, *De revolutionibus (1543), trans. by Owen Gingerich*
Altogether, therefore, thirty-four circles suffice to explain the entire structure of the
universe and the entire ballet of the planets.
Copernicus, *Commentariolus (1510–1514)*

For more than a millennium, astronomy accepted Ptolemy's view of a solar system
controlled by an intricate system of epicyclic curves and their extensions. Copernicus
shared this belief. The quotation above refers to Copernicus's use of epicycles with
34 circles to model the motion of the six known planets. (In later work, he expanded
this number to 48.) Because epicycles were banished from astronomy centuries ago, it
is unimportant today how many circles Copernicus used. Copernicus's great achievement was his understanding that Earth is a planet and that all the planets, including
Earth, orbit about the Sun.

In *Almagest*, Ptolemy explained the motion of each planet by using a three-
dimensional model that agreed rather well with the two-dimensional observational
data available to Ptolemy. Furthermore, Ptolemy's model agrees remarkably well with
three-dimensional planetary motion as it is understood today even though this could
not be verified in Ptolemy's time. Ptolemy's flaw is that he modeled the the planets
separately, each with a different, arbitrarily chosen, scale of distance. Thus, Ptolemy
had no idea of the distance from one planet to another or to the Sun. For example,
Ptolemy was unaware that sometimes Venus and Earth are on opposite sides of
the Sun. (However, later in *Planetary Hypotheses* Ptolemy presents speculative ideas
concerning planetary distances based on the crystalline spheres discussed in Chapter
2—an unfruitful line of thought.)

Copernicus does not tell how he made the great leap from the Ptolemaic system.
Nevertheless, I believe that it is useful to see his transition from the Ptolemaic theory as
a two-step process. The first step is Copernicus's understanding of planetary distances,
and the second step is the placement of the Sun as the fixed center of the solar system.

The first step was later confirmed by Galileo's telescopic observations of the phases of Venus. He claimed to have confirmed the second step, but, as we now know, it is impossible to prove on kinematic grounds alone that the Sun is the motionless center of the solar system. (See Section 3.2.) Confirmation of step two had to wait for Newton's theory of dynamics.

Copernicus's great insight was not based on new experimental evidence or finding flaws in the reasoning of his predecessors. Rather, it was his ability to cast off unnecessary complications of the Ptolemaic theory—his use of Occam's Razor.

6.1. ADJUSTING THE PTOLEMAIC THEORY

Copernicus studied *The Epitome of the Almagest*, a work by the German astronomer and mathematician Regiomontanus (1436–1476). This work mentions that Ptolemy referred briefly in Book 12 of *Almagest* to the possibility of exchanging the deferent and epicycle of an epicyclic motion.[1] This assertion is discussed in Section 3.1 where I have called it the Fundamental Theorem of Epicyclic Motion (Theorem 3.1). This theorem is used below to show a logical transition from the Ptolemaic theory to the Copernican. It is possible that Copernicus followed this pattern of thought, but this is not known. Nevertheless, I believe that this two-step analysis shows the logical structure of the Copernican theory.

In *Almagest*, Ptolemy treats each planet without any connection to the other planets. It is possible to diagram all of the planets together, each moving separately as Ptolemy claims. Figure 6.1 is such a diagram, showing the Ptolemaic deferent-epicycle model for each planet, but it uses a different arbitrarily chosen scale of distance for each deferent-epicycle pair.

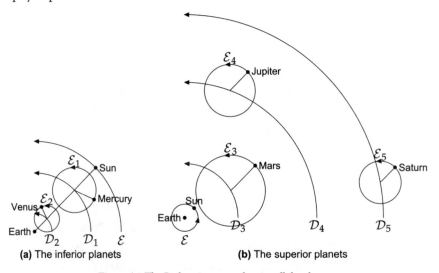

(a) The inferior planets   **(b)** The superior planets

Figure 6.1 The Ptolemaic system showing all the planets.

Figure 6.1 is divided into two subfigures: 6.1a showing the Sun and the inferior planets, Mercury and Venus, and 6.1b showing the Sun and the superior planets, Mars, Jupiter, and Saturn. The subfigure 6.1a is shown at a higher level of magnification so that $\mathcal{E}$ represents the path of the Sun in both figures. The deferent circles are $\mathcal{D}_1 \ldots \mathcal{D}_5$ and the epicycles are $\mathcal{E}_1 \ldots \mathcal{E}_5$.

The Ptolemaic system has dimensions that are unspecified

In Figure 6.1, certain relationships are specified by Ptolemy's model, but other dimensions are arbitrary. The following three conditions are required:

1. For each planet, the ratio of the radii of the corresponding deferent–epicycle pair is specified by Ptolemy's model. These ratios are used in the construction of Figure 6.1. For example, for Mars this ratio—the ratio of the radius of $\mathcal{D}_3$ to the radius of $\mathcal{E}_3$—is equal to 1.524. It is shown in Section 4.3 how Ptolemy determined this ratio for Mars.

2. For all the planets, the rates of rotation of their deferents and epicycles about their centers are known. In particular, the deferents for Mercury and Mars ($\mathcal{D}_1$ and $\mathcal{D}_2$), the epicycles of Mars, Jupiter, and Neptune($\mathcal{E}_3$, $\mathcal{E}_4$, and $\mathcal{E}_5$), and the Sun ($\mathcal{E}$) have rotation rates of one revolution per year.

3. In Figure 6.1b, the Earth–Sun direction must be parallel to the radii connecting each superior planet to the center of its epicycle. This can be seen as follows. Ptolemy's deferent–epicycle model for Mars is shown in Figure 3.2. In this figure, OM is a radius connecting the center of epicycle O to the planet Mars located at M. In Ptolemy's model, the rate of rotation of a superior planet about its epicycle, in this case the rate of rotation of M about O, is equal to one revolution per year. Since $\overrightarrow{MO}$ points to the Sun at opposition—as discussed in Section 4.3—it follows that $\overrightarrow{MO}$ remains, as claimed, parallel to the Earth–Sun direction.

On the other hand, the sizes of the deferents ($\mathcal{D}_i$, $i = 1, \ldots, 5$) are arbitrary, so long as the ratios specified above in 1 have the correct values.

To adjust Ptolemy's geocentric theory, two steps are needed. First, exchange the deferents and epicycles of the superior planets. Second, adjust the sizes of the deferents and epicycles while keeping constant the ratios mentioned above, so that the new deferents all coincide with $\mathcal{E}$, the path of the Sun.

Exchanging deferents and epicycles of the superior planets

According to Theorem 3.1 in Section 3.1, the Fundamental Theorem of Epicyclic Motion, exchanging deferents and epicycles in Figure 6.1b does not change any of the

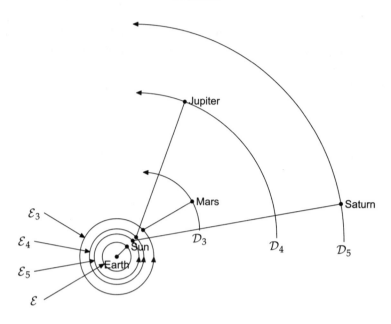

Figure 6.2 Modification of Figure 6.1b, exchanging deferents and epicycles of the superior planets.

epicyclic motions. Figure 6.2 shows the result of making this exchange. Notice that the new deferents, $\mathcal{E}_3$, $\mathcal{E}_4$, and $\mathcal{E}_5$, are concentric circles about Earth, together with $\mathcal{E}$, the path of the Sun.

Adjusting the sizes of deferents and epicycles

The Ptolemaic system permits magnifying or diminishing each deferent–epicycle pair in Figure 6.1a or 6.1b by a different factor while keeping the Earth in its fixed position. Because the newly defined deferents $\mathcal{D}_1$, $\mathcal{D}_2$, $\mathcal{E}_3$, $\mathcal{E}_4$ and $\mathcal{E}_5$ have the same rotation rate as the Sun ($\mathcal{E}$), apply a suitable factor to each deferent-epicycle pair so that *the deferents are all identical with $\mathcal{E}$,* the circular path of the Sun. Applying this transformation to Figure 6.1b results in the much simpler representation of the superior planets shown in Figure 6.3b.

The transformation of Figure 6.1a showing the motion of the inferior planets is simpler because it is not necessary to invoke Theorem 3.1. Figure 6.3a shows the similar transformation so that the deferents of Mercury ($\mathcal{D}_1$) and Venus ($\mathcal{D}_2$) are also identical with the circle $\mathcal{E}$. Thus, in Figures 6.3a and 6.3b, $\mathcal{E}$ serves as a deferent for all of the planets. This transformation could be justified on the grounds that the result is a great simplification. It represents an application of Occam's Razor to the more complex version of the Ptolemaic system shown in Figure 6.1.

Furthermore, Figure 6.3 has a simpler kinematic interpretation than Figure 6.1. In fact, Figure 6.3 can be interpreted as orbital motion of the planets around the Sun,

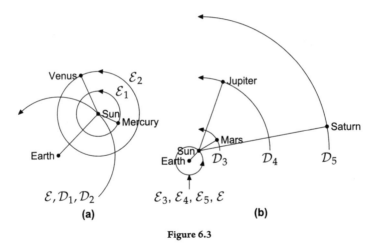

Figure 6.3

which, in turn, orbits about the Earth. This interpretation was adopted by Tycho Brahe and other post-Copernican astronomers. These astronomers adopted Copernicus's theory of the relative distances of the solar system but were unwilling to accept Copernicus's heliocentric theory.

In the time of Copernicus, this one-deferent version of the Ptolemaic system was not supported by any observational evidence. Indeed, such confirmation is beyond the capabilities of naked-eye observation. Galileo (1564–1642) provided the first confirmation—discussed further in Section 6.3—with his telescopic observation of the phases of Venus. However, Copernicus noted—see the epigraph at the start of this chapter—the beautiful connection between the sizes and the periods of the orbits of the planets about the Sun. This connection was made quantitative later by Kepler.

Placing the Sun at the center of the solar system

Although step two of Copernicus's theory, putting the Sun at the center of the solar system, is, from a purely mathematical perspective, a minor extension of the one-deferent extension of the Ptolemaic theory, it is amply justified because it eliminates the unduly complicated deferent–epicycle model, and it prepares the way for the Newtonian theory of universal gravitation.

If the motion depicted in Figure 6.3 is *viewed relative to the Sun*, the result is Figure 3.7, the orrery model in which the planets have circular orbits centered at the Sun. Copernicus tweaked this system by adding epicyclets to improve its agreement with observational data.

6.2. COPERNICUS

The Polish astronomer Nicolaus Copernicus is renowned as the instigator of a revolution in astronomy (see Kuhn, 1985), but revolution was not his intention. Copernicus wanted to repair what he perceived as a flaw in the Ptolemaic system. He was generally satisfied with Ptolemy's epicycles. However, he wished to avoid Ptolemy's equant (Section 4.5) because he considered it a departure from Plato's dictum that planetary motion must be based on uniform circular motion.

Copernicus did not intend to reduce any predictive inaccuracies in the Ptolemaic system, nor did he do so. Copernicus used Ptolemy's astronomical data. He was more interested in theory than observation.

Copernicus was not the first to propose a heliocentric model of the solar system. That honor goes to Aristarchus of Samos, eighteen centuries before Copernicus, although Aristarchus did not win the consensus of his colleagues.

Copernicus very cautiously withheld publishing his theories until he was urged to do so near the end of his life. On the other hand, Galileo, among others, aggressively and fearlessly (some say tactlessly) supported Copernicus's heliocentric theory, despite suspicion of heresy leading to house arrest imposed by the Roman Catholic Church.

Copernicus was canon at the cathedral of the Polish Baltic coastal town of Frauenburg (now Frombork). Copernicus held his position as canon for the period from 1497 till the end of his life in 1543. This position provided him with an income and enough free time to pursue his interest in astronomy, which resulted in his major work *De revolutionibus orbium cœlestium,* "On the Revolutions of the Heavenly Spheres," largely completed in the period 1512–1530. Near the end of his life in 1543 Copernicus reluctantly agreed to publication. No doubt fearing disapproval from the Roman Catholic Church, he agreed to publication only at the urging of his friends Rheticus and Giese.

De revolutionibus contains a preface, written by Andreas Osiander, that claims that the work is simply computational artifice containing no assertion that the Sun is truly the center of the solar system. The following is a translation of this preface:[2]

> Since the novelty of the hypotheses of this work has already been widely reported, I have no doubt that some learned men have taken serious offense because the book declares that the Earth moves; these men undoubtedly believe that the long established liberal arts should not be thrown into confusion. But if they examine the matter closely, they will find that the author of this work has done nothing blameworthy. For it is the duty of an astronomer to record celestial motions through careful observation. Then, turning to the causes of these motions he must conceive and devise hypotheses about them, since he cannot in any way attain to the true cause.... The present author has performed both these duties excellently. For these hypotheses need not be true nor even probable; if they provide a calculus consistent with the observations, that alone is sufficient.... Now when there are offered for the same motion different hypotheses, the astronomer will accept the one which is the easiest to grasp ... let no one expect anything certain from astronomy,

which cannot furnish it, lest he accept as the truth ideas conceived for another purpose, and depart from this study a greater fool than when he entered it. Farewell.

Osiander's motive seems to be to forestall criticism from religious authorities. On the other hand, it is correct to say that the principal benefit of the Copernican heliocentric theory is simplicity rather than truth. Indeed, the motion of the solar system can be correctly charted with reference to any coordinate system. Of course, this does not in any way diminish the importance of the Copernican revolution.

In the period 1510–1514, Copernicus wrote, but did not publish, a short preliminary announcement titled *Commentariolus* including a list of seven axioms:

1. The celestial bodies do not all move around the same center.
2. Earth's center is not the center of the universe.
3. The center of the universe is near the Sun. [Copernicus placed the center of rotation a small distance away from the Sun to give his scheme of epicyclic curves best agreement with observation.]
4. The distance from Earth to the Sun is imperceptible compared with the distance to the stars.
5. The rotation of Earth accounts for the apparent daily rotation of the stars.
6. The annual motion of the Sun against the fixed stars is due to Earth's orbit about the Sun.
7. The apparent retrograde motion of the planets is caused by the motion of Earth and the planets about the Sun.

In *De revolutionibus*, Copernicus developed these assertions fully. Techniques for computing some of the astronomical velocities and distances used in Copernicus's model were known already to ancient astronomers for use in the deferent-epicycle model. However, Copernicus was the first to understand that these distances and velocities related to the planets directly rather than to hypothetical constructions (the epicycles and deferents).

Copernicus was a conservative, not a revolutionary. It is likely that he believed that the Sun was truly the center of the solar system, but his stated reason for finding fault with Ptolemy's geocentric astronomy was that it departed from Plato's dictum to use uniform circular motion. In particular, he objected to Ptolemy's use of the equant, described in Section 4.5. He wished to create a theory of the solar system that used epicyclic curves, deferents, and eccentrics but no equants—a goal that his heliocentric system achieved. Copernicus used epicyclic curves freely. His system was just as complex and no more accurate than the Ptolemaic system in use at that time.

Nevertheless, Copernicus ignited a revolution. Galileo and others followed the Copernican theory, not because it avoided the use of equants, but because it placed the Sun at the center of the solar system.

6.3. GALILEO

Galileo Galilei contributed to two sciences—astronomy and mechanics. These contributions are treated in his two dialogues—*Dialogue Concerning the Two Chief Systems of the World—Ptolemaic and Copernican* (1632), and *Discourses and Mathematical Demonstrations Concerning the Two New Sciences* (1638). These two new sciences are branches of mechanics—*strength of materials* and *dynamics*. Galileo's contributions to mechanics are discussed in Section 9.2.

In 1609, Galileo learned about the recent invention, the telescope, and decided to construct one of his own. He used his telescope to explore the heavens. He quickly made disturbing discoveries that contradicted many of the old ideas about astronomy. He upset the notion of the cool simplicity and perfection of the heavens. Mysteries, complexities, and blemishes abound. The Sun has spots; the Moon has craters, mountains, and valleys; Jupiter has moons; Venus has phases like the Moon; Saturn has "ears"—no, they must be rings.

Galileo proposed that precise timing of eclipses of the moons of Jupiter could enable mariners to determine longitude—an urgent navigation problem of that time. Galileo invented devices for this purpose, but none achieved practical usage. In the century after Galileo, accurate clocks together with celestial observation using a sextant sufficed for determining longitude.

Galileo advocated the Copernican theory publicly. Despite an order, in 1616, from the Inquisition banning him from teaching the Copernican theory, he published *Systems of the World* (see Figure 6.4). His book was written, not timidly in Latin for scholars only, but boldly in Italian for the general reader. Galileo claimed that his book was only a hypothetical discussion; nevertheless, his fictional Copernican spokesman Salviati clearly defeated the arguments presented to the layman Sagredo by the Aristotelian Simplicio. For his alleged disobedience, Galileo was tried in 1633, charged with "vehement suspicion of heresy." He was convicted of disobeying orders forbidding him to teach heliocentric cosmology and given a sentence of life imprisonment—which he served in the form of house arrest until his death in 1642. He was forbidden to publish—but publish he did. *Two New Sciences* was published in Leiden, Holland, out of reach of the Inquisition. Galileo was not punished additionally for this infraction.

Outrage at Galileo's trial has given fame to Galileo's astronomical work, but many believe that his work in mechanics, the topic of *Two New Sciences*, was his greatest

Figure 6.4 Frontispiece of Galileo's *Dialogue Concerning the Two Chief Systems of the World* (1632).

achievement. Galileo pioneered the theory of falling bodies—leading to Newton's laws of dynamics and theory of universal gravitation. Thus, Galileo advanced two threads of thought that would be taken up a century later by Newton who wove them into the first theory that accounted for all aspects of planetary motion.

Galileo may have had a role in converting Johannes Kepler to the heliocentric premise. It was Kepler who finally brought the astronomical revolution to a conclusion. Galileo supported Copernicus's heliocentric model. He asserted that it is fact, not just computational expediency, that Earth and the planets orbit about the Sun. His support of the heliocentric system provoked the opposition of the Roman Catholic Church, and this controversy stimulated the need to settle the issue with more careful astronomical measurements.

The phases of Venus

Galileo made telescopic observations that he claimed gave support to the Copernican theory. In particular, he discovered that Venus, like the Moon, has crescent phases, which he explained by the rotation of Venus about the Sun. In a letter to the Tuscan ambassador to Prague, Galileo secreted information concerning the phases of Venus in an anagram.

Galileo's drawings of Venus are shown in Figure 6.5a, showing that when Venus is in its narrowest crescent phase 1; the "new" phase) its diameter is about six times greater than in its "full" phase (5). He believed that this observation gave strong support to the heliocentric theory of the solar system. This is illustrated in Figure 6.5b, which shows relative positions of Venus and Earth corresponding to the phases shown in Figure 6.5a. The most distant position of Venus from Earth corresponds to its full phase.

The ratio of the farthest to nearest distances between Earth's and Venus's orbits is known to be 6.2280. This is a good confirmation of the accuracy of Galileo's observations because in Figure 6.5a the diameter of phase 1 is shown to be about six times as great as the diameter of phase 5. Ptolemy, since he had no telescope, used a

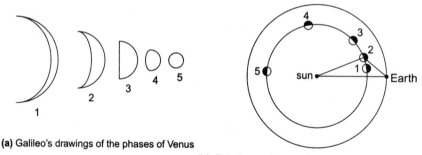

(a) Galileo's drawings of the phases of Venus

(b) Relative positions of Venus and Earth corresponding to the phases observed by Galileo

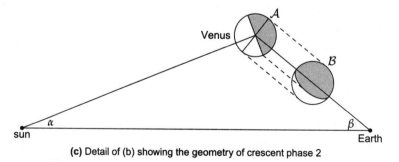

(c) Detail of (b) showing the geometry of crescent phase 2

Figure 6.5 Galileo's telescopic observations of the phases of Venus supported the heliocentric theory of the solar system. In (c), Venus is greatly magnified. The relationship is shown between the Sun and the positions of Venus and Earth. An orthogonal projection shows the crescent shape of phase 2.

different method of calculation involving the diameters of the deferent and epicycle of Venus. His remarkably accurate estimate of the ratio was 6.5.

Figure 6.5c is an enlargement of a detail in Figure 6.5b consisting of the triangle formed by the Sun, Earth, and image 2 of Venus, showing the geometry underlying the observations of the phases of Venus. A is Venus, greatly enlarged, image 2 in Figure 6.5b. B is an orthogonal projection of A (a rotation of A by 90°) showing the crescent shape seen from Earth. The angle β represents the apparent distance in the sky between the Sun and Venus. If β, is equal to zero or 180°, Venus is obscured by the glare of the Sun.[3]

Galileo's observations did not *prove* the heliocentric theory of the solar system. They are explained equally well by using a geocentric model.[4] The story of Earthship 2 in Section 3.2 shows that the difference between the heliocentric and geocentric theory can never be more than a relative point of view, but Galileo did demonstrate that the geocentric premise was unnecessarily complicated and of little use in any systematic treatment of astronomy. The coming Newtonian theory of dynamics would demonstrate this more convincingly.

Galileo's observations with his telescope hastened the end of the dogma that the heavens are a region of eternal and immutable perfection. Galileo found pimples and moles on those heavenly bodies—spots on the Sun, craters on the Moon. Today we are comfortable with the idea that Earth is a minute part of an incredibly complex cosmos, but this idea was upsetting in Galileo's time.

The hearing impaired—I am one of them—do not always welcome a hearing aid. The peace of a gentle ringing in the ears is replaced by a buzzing confusion of traffic sounds, half-heard conversations, and clattering dishes. Galileo augmented our senses, and the universe can never be as it was.

The work of Copernicus and Galileo set the stage for Kepler's discovery that planetary orbits are elliptical. In preparation, the next chapter discusses abstract properties of the ellipse.

7

CIRCLES NO MORE

On a cloth untrue
With a twisted cue
And elliptical billiard balls.
W. S. Gilbert, *The Mikado* (*1885*)

Early in the seventeenth century, Kepler discovered that planetary orbits are ellipses. In preparation for this story, to be told in the next chapter, we now discuss some properties of these curves.

7.1. THE ELLIPSE

Apollonius of Perga, a mathematician of the Alexandrian school, studied the ellipse together with the parabola and the hyperbola—collectively known as the sections. He gave these curves the names by which they are known today. On this subject, Apollonius wrote the treatise *Conics*, consisting of eight books, of which seven survive— the first four in the original Greek, the fifth through seventh in Arabic translation. The seven that survive contain no less than 387 propositions. Nevertheless, Apollonius did not suggest the importance in astronomy of the ellipse and the other conics. As mentioned earlier, he proposed epicyclic curves to describe the motion of the planets.

All the conic sections—ellipse, parabola, and hyperbola—can be seen as the boundary curves between light and dark when shining a flashlight on a flat wall in a darkened room.

An ellipse is an oval curve of a very special sort, to be defined shortly. A photograph of my desk top would represent the circular lip of my coffee cup as an ellipse. We see hundreds of ellipses each day because that is how the eye perceives an oblique view of a circle—indeed, the circle itself is a special kind of ellipse.

The above observation could be formalized into a definition, but there is a quicker route to the astronomical applications of the ellipse, shown in Figure 7.1a. The construction takes place in a fixed two-dimensional plane. An ellipse $\mathcal{E}$ is determined by choosing points E and F, called the *foci*, together with a number d larger than the

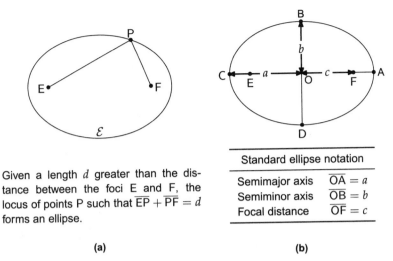

(a)

Given a length d greater than the distance between the foci E and F, the locus of points P such that $\overline{EP} + \overline{PF} = d$ forms an ellipse.

Standard ellipse notation	
Semimajor axis	$\overline{OA} = a$
Semiminor axis	$\overline{OB} = b$
Focal distance	$\overline{OF} = c$

(b)

Figure 7.1 The ellipse.

distance between E and F. The ellipse $\mathcal{E}$ is defined as the locus[1] of points P such that the sum of the two distances $\overline{PE}$ and $\overline{PF}$ is equal to d —that is, such that $\overline{EP} + \overline{PF} = d$. If points E and F happen to coincide, the ellipse becomes a circle of radius $d/2$.

To verify the correctness of Figure 7.1a, fasten two pins at E and F. (To avoid defacing this book, one could make a photocopy of Figure 7.1a.) Fasten a thread of length d to the two pins. Use a pencil to hold the thread taut, and slide the pencil against the taut thread, thereby drawing a curve that should coincide with Figure 7.1a.

Figure 7.1b shows some standard terminology for the ellipse. Line segment AC passes through E and F, the two foci of the ellipse. Point O, the center of the ellipse, is the midpoint of AC. The segment BD is drawn perpendicular to AB through O. The lengths $\overline{OA} = a$, $\overline{OB} = b$, and $\overline{OF} = c$, are called, respectively, the semimajor axis, the semiminor axis, and the focal distance. In Figure 7.1b, the notation A, B, ..., F and O is arbitrary, but the use of the letters a, b, and c is standard.

Putting P equal to A shows that d, the length of the string, is equal to $2a$. Putting P equal to B and using the Pythagorean theorem establishes the equation $a^2 = b^2 + c^2$.

The reflection property

It will be shown below that the ellipse has the following remarkable property. Suppose the boundary of the ellipse is a mirror. Then a ray of light emanating from a point source at one focus passes through the other focus. For example, in Figure 7.1a, EP is a light ray emanating at E, and PF is the reflected ray passing through F.

The reflection property is stated in the following theorem, which refers to Figure 7.2.

Proposition 7.1. *Let P be a point on an ellipse $\mathcal{E}$ with foci E and F. A line $\mathcal{T}$ containing P is tangent to $\mathcal{E}$ if and only if the angles $\angle APE$ and $\angle BPF$ are equal.*

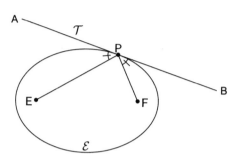

Figure 7.2 Ellipse $\mathcal{E}$ with tangent line $\mathcal{T}$.

Sound waves are capable of a similar focusing effect. Imagine a room in the shape of an ellipse with walls that reflect sound. Then, regardless of the size of the room, any sound that emanates from one focus of the ellipse can be heard with astonishing clarity at the other focus. A room of this sort is called a *whispering gallery*. Usually the elliptical nature of the room is partial, subtle, and unintended. Notable whispering galleries exist at Statuary Hall in the U.S. National Capitol and at Saint Paul's Cathedral in London.

The ellipse is the boundary between the *interior* and *exterior* regions. A point Q belongs to the interior or exterior region or to the ellipse itself depending on whether

$$\overline{EQ} + \overline{FQ} < d,$$

$$\overline{EQ} + \overline{FQ} > d, \text{ or}$$

$$\overline{EQ} + \overline{FQ} = d$$

The following proof uses concepts that deserve preliminary mention:

1. A *tangent* line to an ellipse is characterized as follows: A single point on the line belongs to the ellipse, and all other points on the line are exterior to the ellipse.
2. *Reflection across a line.* In Figure 7.3a, point G is the reflection of F across the line $\mathcal{T}$. Note that $\mathcal{T}$ is the perpendicular bisector of the line segment FG.
3. The proof depends on a geometric fact that has turned into a cliché in everyday speech: "A straight line is the shortest distance between two points." Specifically, the proof makes crucial use of the proposition that the length of a side of a triangle must be less than the sum of the lengths of the other two sides.

As shown in Figure 7.3b, we ignore, for the moment, the ellipse $\mathcal{E}$ and consider a line $\mathcal{T}$ through the point P on the boundary of the ellipse such that the angles $\angle QPE$ and $\angle RPF$ are equal. Theorem 7.1 is equivalent to the following proposition.

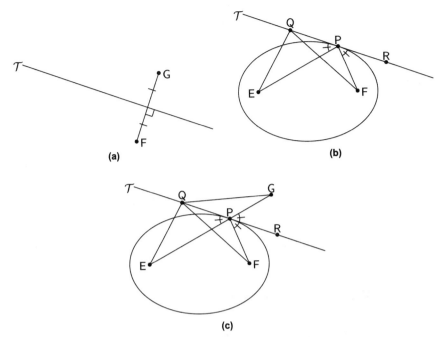

Figure 7.3 Proof of Proposition 7.2. The labeling of these three figures and Figure 7.2 is consistent. For example, F represents the same point in all four figures.

Proposition 7.2. *Referring to Figure 7.3b, let Q be an arbitrary point, different from P, on the line $\mathcal{T}$. Then the sum of the distances $\overline{QE}$ and $\overline{QF}$ is greater than d, the sum of $\overline{PE}$ and $\overline{PF}$.*

Proof. Referring to Figure 7.3c, consider the reflection of triangle $\triangle FPQ$ across line $\mathcal{T}$. The image of triangle $\triangle FPQ$ is triangle $\triangle GPQ$, and angle $\angle RPG$ is equal to angle $\angle RPF$, which was constructed equal to angle $\angle QPE$. Thus $\angle QPE$ is equal to $\angle RPG$, which implies that EPG is a straight line. In fact, the angles $\angle QPE$ and $\angle RPG$ are equal vertical angles of the intersecting lines $\mathcal{T}$ and EPG. Since the length $\overline{PG}$ is equal to $\overline{PF}$, the length of the line segment EPG is equal to $\overline{PE} + \overline{PF}$, which, from the definition of ellipse $\mathcal{E}$, is equal to d.

Now $\overline{QE} + \overline{QG}$ is greater than $\overline{EPG} = d$ because the length of this side of triangle EQG must be less than the sum of the lengths of the other two sides— that is, $\overline{QE} + \overline{QG}$. Since $\overline{QG} = \overline{QF}$, it follows, as claimed, that $\overline{QE} + \overline{QF}$ is greater than d. $\qquad\square$

Thus $\overline{QE} + \overline{QF} \geq d$ for an arbitrary point Q on $\mathcal{T}$, and equality holds only if Q = P. It follows that, while P belongs to the ellipse $\mathcal{E}$, every other point of line $\mathcal{T}$ belongs to the *exterior* of the ellipse. This means line $\mathcal{T}$—shown in Figures 7.3b, 7.3c, and 7.4a—is *tangent* to the ellipse and P is the point of tangency.

Constructing tangents

This section discusses solutions to two problems concerning tangents to an ellipse.

Problem 7.1. Draw the tangent through a given point on the ellipse.

Problem 7.2. Draw the two tangents having a given direction.

The solutions are facilitated by the use of two auxiliary circles—the *focal* circle for Problem 7.1 and the *major* circle for Problem 7.2.

Referring to Figure 7.4a, suppose $\mathcal{E}$ is a given ellipse with foci E and F. Furthermore, suppose P is an arbitrary point on the ellipse such that $\overline{PE} + \overline{PF} = d$. Then the previous discussion makes it is possible to construct the tangent $\mathcal{T}$ containing the given point P:

1. Locate point G by extending the line EP such that the length of the line segment EPG is equal to $d = 2a$.
2. Point S denotes the midpoint of FG. The perpendicular bisector of the segment FSG is $\mathcal{T}$, the required tangent of $\mathcal{E}$.

The perpendicular bisector of a given line segment is one of the classic ruler-and-compass geometrical constructions. Thus, given the foci E and F and a point P on the ellipse, it is possible to construct, using ruler and compass only, a line tangent to the ellipse at point P.

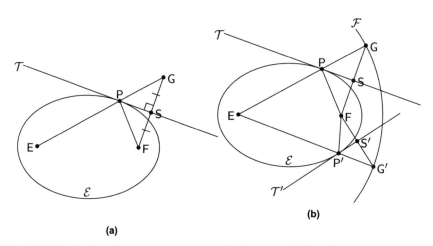

Figure 7.4 Construction of tangents to an ellipse. In (b), the circle $\mathcal{F}$ of radius $d = 2a$ centered at E is called the *focal* circle. The lines $\mathcal{T}$ and $\mathcal{T}'$ are tangent to the ellipse $\mathcal{E}$ at points P and P', respectively.

The focal circle

Figure 7.4b shows the auxiliary circle $\mathcal{F}$, the *focal* circle, facilitating the solution of Problem 7.1. This figure shows the construction of the tangent $\mathcal{T}$ at point P, as in Figure 7.4a, as well as the tangent at an additional point P'. To find the tangent $\mathcal{T}'$ at the second point P' on the ellipse $\mathcal{E}$, draw the focal circle $\mathcal{F}$ centered at E with radius d. Extend the line segment EP' to find its point of intersection G' with circle $\mathcal{F}$. The tangent $\mathcal{T}'$ is the perpendicular bisector of the line segment FG'.

The major circle

This section discusses a further method for finding tangents to an ellipse—the construction of a tangent line with a prescribed direction.

The *major* circle of an ellipse is the circle centered at the center of the ellipse with a radius equal to the semimajor axis—for example, $\mathcal{M}$ in Figure 7.5a. Inscribe the rectangle STUV in the circle $\mathcal{M}$ as follows. Draw an arbitrary nonhorizontal line through focus F, intersecting $\mathcal{M}$ at points S and T. Segment UV is obtained by rotating ST 180° about the center O, completing the rectangle STUV. From the symmetry of this construction, the segments FS and FT are equal in length, respectively, to UE and EV.

The rectangle STUV has a couple of remarkable properties detailed in Propositions 7.3 and 7.4. The statement of the proposition refers to Figure 7.5a. Figure 7.5b contains additional constructions that will be used in the proof.

Proposition 7.3. *The lines VS and UT are tangent to the ellipse $\mathcal{E}$.*[2]

Proof. Figure 7.5b contains additions to Figure 7.5a. $\mathcal{F}$ is an arc of the focal circle centered at E with radius $2a$, where a is the semimajor axis of the ellipse $\mathcal{E}$. The extension of line segment ST intersects $\mathcal{F}$ at points G and H. The line segments EG and EH intersect the ellipse at points P and Q, respectively. The line segments FP and PG have equal lengths—in fact, both have a length equal to $2a - \overline{\text{EP}}$. The line segment PS is perpendicular to FS because STUV is a rectangle. Moreover, PS is the altitude of the isosceles triangle $\triangle$FPG. It follows that SP is the perpendicular bisector of FG. From the discussion of the focal circle in the previous section, it follows that SV is tangent to the ellipse $\mathcal{E}$ at point P, and, similarly, TU is tangent to $\mathcal{E}$ at point Q. $\qquad\square$

Proposition 7.3 solves Problem 7.2, to construct tangents with a prescribed direction, apart from the trivial case in which the prescribed direction is vertical, as follows. In Figure 7.5a, choose the line segment SFT *perpendicular* to the prescribed direction. Then SV and TU are the desired tangents to the ellipse $\mathcal{E}$.

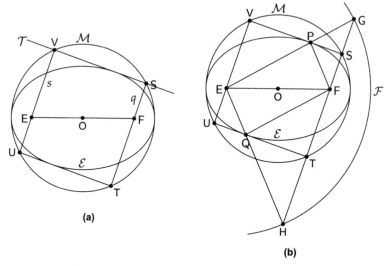

Figure 7.5 Major circle $\mathcal{M}$ with inscribed rectangle STUV.

The following proposition, which refers again to Figure 7.5a, will be used later in the proof of Proposition 10.4.

Proposition 7.4. *Let $\mathcal{T}$ be a tangent to an ellipse $\mathcal{E}$ with foci E and F. Let s and q be the distances from the foci E and F, respectively, to the tangent line $\mathcal{T}$. Then the product sq is equal to b^2, where b is the semiminor axis of the ellipse.*

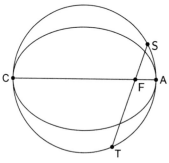

Figure 7.6 Proof of Proposition 7.4.

The proof of this proposition follows from a standard theorem of plane geometry:

Theorem 7.1. *If two chords in a circle intersect, the products of the lengths of their segments are equal.*

Proof of Proposition 7.4. Referring to Figure 7.5a, since $s = \overline{EV} = \overline{TF}$, it is sufficient to show $\overline{TF} \cdot \overline{SF} = b^2$. Referring to Figure 7.6, according to Theorem 7.1,

$$\overline{SF} \cdot \overline{FT} = \overline{AF} \cdot \overline{FC} \tag{7.1}$$

But $\overline{AF} = a - c$ and $\overline{FC} = a + c$ where a is the semimajor axis and c is the focal distance. Thus, the right-hand side of the equation (7.1) is equal to

$$(a + c)(a - c) = a^2 - c^2 = b^2$$

where b is the semiminor axis of the ellipse. $\qquad\square$

The following section discusses, without proofs, two beautiful properties of the ellipse.

7.2. TWO PEARLS

Pascal's Theorem

The French mathematician Blaise Pascal (1623–1662) was a child prodigy. He was born to a wealthy French family. His father, Étienne Pascal, who carefully planned Blaise's education, decided that his son should have no contact with mathematics before the age of fifteen and, to this end, removed all mathematical books from his house. However, his son disobeyed his wishes and rediscovered some theorems of Euclid at age twelve. His father relented, and Blaise soon had contact with eminent mathematicians of the time, including Girard Desargues (1591–1661) who pioneered the field of *projective geometry,* an extension of Euclidean plane geometry. At age sixteen, Pascal discovered a pearl of projective geometry. He called it "the mystic hexagram."

Pascal discovered a remarkable property of a hexagon inscribed in an ellipse. Any ordering of six points on an ellipse forms a suitable hexagon, even if the sides intersect as in Figure 7.7a. Pascal's theorem concerns the *opposite* sides of such a hexagon. In Figure 7.7a, the three pairs of opposite sides are (1) AB and DE, (2) BC and EF, and (3) CD and FA.

Theorem 7.2 (Pascal's Mystic Hexagram). *For any hexagon inscribed in an ellipse, let P, Q, and R be points of intersection of each of the three pairs of opposite sides. Then P, Q, and R are collinear.*

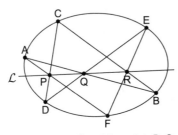
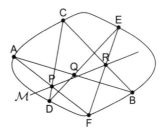

(a) Ellipse. Line $\mathcal{L}$ contains points P, Q, and R.

(b) Not an ellipse. Line $\mathcal{M}$ contains points P and Q but misses point Q.

Figure 7.7 Pascal's theorem. Opposite sides of a hexagon inscribed in an ellipse must intersect in collinear points. In the above figures, opposite sides of hexagon ABCDEF intersect as follows: (1) AB and DE intersect at Q, (2) BC and EF intersect at R, and (3) CD and FA intersect at P. The theorem asserts that if the curve is an ellipse, as in (a) P, Q, and R are collinear because line $\mathcal{L}$ contains these three points. Since (b) is not an ellipse, Pascal's theorem does not apply, and, in fact, P, Q, and R are not collinear.

For example, in Figure 7.7a, the intersection points P, Q, and R are collinear because they lie on line $\mathcal{L}$.

Figure 7.7b shows that if the curve is not an ellipse, then the intersection points P, Q, and R need not be collinear. Pascal's theorem remains true if the ellipse is replaced by any conic section—hyperbola, parabola, or even a pair of straight lines.

The converse of Pascal's theorem is true in the following sense. If a curve has the property that for *every* inscribed hexagon the intersections of opposite sides are collinear, then the curve must be a conic section—an ellipse, parabola, hyperbola—or a pair of straight lines (a degenerate conic section).

7.2.1 Brianchon's theorem

Duality is a concept that sets projective geometry apart from classical plane Euclidean geometry. Roughly, any true general statement in projective geometry remains true if the words "point" and "line" are (judiciously) interchanged. Points on a curve become tangents to the curve. Collinear points become concurrent lines. Following this principle, one would expect that Pascal's theorem could be so transformed. However, it took a long time—almost two hundred years—for the other shoe to drop. The dual of Pascal's theorem was enunciated in 1806 by a young French mathematician named Charles Julien Brianchon (1783–1864).

Here is how Pascal's theorem must be altered to become Brianchon's theorem:

(~~Pascal's Mystic Hexagram~~ Brianchon's theorem). For any hexagon ~~inscribed in~~ *circumscribed about* an ellipse, let ~~P, Q, and R~~ $\mathcal{L}$, $\mathcal{M}$, and $\mathcal{N}$ be ~~points of intersection of lines containing~~ each of the three pairs of opposite ~~sides~~ *vertices*. Then ~~P, Q, and R~~ $\mathcal{L}$, $\mathcal{M}$, and $\mathcal{N}$ are ~~collinear~~ *concurrent*. Cleaning this up, we have the following:

Theorem 7.3 (Brianchon's theorem). *For any hexagon circumscribed about an ellipse, let* $\mathcal{L}$, $\mathcal{M}$, *and* $\mathcal{N}$ *be lines containing each of the three pairs of opposite vertices. Then the lines* $\mathcal{L}$, $\mathcal{M}$, *and* $\mathcal{N}$ *are concurrent.*

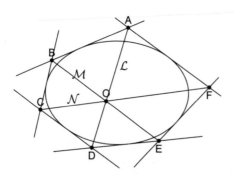

Figure 7.8 Brianchon's theorem.

Figure 7.8 confirms Brianchon's theorem. The curve is an ellipse, and, indeed, the lines $\mathcal{L}$, $\mathcal{M}$, and $\mathcal{N}$ are concurrent in point O.

Infinity

A regular hexagon inscribed in a circle, shown in Figure 7.9, appears to be an exception to Pascal's theorem because

Figure 7.9

opposite sides—for example, AF and CD—are parallel and extensions of these lines do not intersect. Projective geometry takes care of this by defining a *line of points at infinity.* Each ordinary straight line is augmented with a point at infinity. By definition, parallel lines share the same point at infinity, and nonparallel lines have distinct points at infinity. Thus, each of the three pairs of opposite sides of the regular hexagon in Figure 7.9 intersect in distinct points on the line at infinity. In other words, the line at infinity plays the role of line $\mathcal{L}$ in Figure 7.7a.

Infinity in projective geometry is an intellectual construction, not a speculation. It is introduced, not to jibe with "reality," but to provide a satisfying completion to a geometrical theory. Infinity in mathematics has nothing to do with awe at the immense size of the universe.

Mathematics modifies the concept of infinity to suit the geometrical context. There is a different theory of the two-dimensional plane, the *complex plane,* in which there is just *one* point at infinity. No confusion arises. The complex plane does not interfere with the projective plane.

7.3. TRACKING PLANETS

The most difficult problem for astronomical pioneers was to deduce the three-dimensional position of a planet from two-dimensional observations such as azimuth and elevation (see Section 2.2). This problem will be discussed in the next chapter in Section 8.2. This section assumes that this problem has been solved and discusses the following problem:

Problem. Suppose that we have found the three-dimensional coordinates for a small number of observations of a planet. Could the path be a circle or an ellipse? If so, specify the possible circle or ellipse.

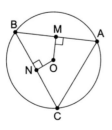

Figure 7.10 Circumcircle.

Of course, neither alternative is possible unless the observed points lie in a plane. There are sophisticated statistical tools that take account of errors of observation. On a simpler level, the following two propositions provide basic solutions to the problem.

Proposition 7.5. *Given three noncollinear points, there exists exactly one circle containing them.*

Figure 7.10 shows this geometrical construction. The triangle △ABC determines a circumscribed circle, called the *circumcircle*. The center O of the circumcircle is the intersection of the perpendicular bisectors of any two sides of triangle △ABC.

Proposition 7.6.   *Given five points in a plane, no three of them collinear, there exists exactly one conic section (ellipse, parabola, or hyperbola) containing them.*

In fact, given five such points, every other point of the conic section can be constructed geometrically by using Pascal's mystic hexagram (Figure 7.7a). This construction is shown, step by step, in Figure 7.11. The five given points are A, B, C, D, and E. The following gives a geometrical construction to locate a sixth point F such that ABCDEF is a mystic hexagram.

The first step (Figure 7.11a) is to draw as much of the hexagram as possible omitting the unknown vertex F. The result is the broken line ABCDE. When the hexagram is complete, AB and DE will be opposite sides, and we denote their intersection Q.

The second step (Figure 7.11b) is to draw an *arbitrary* line $\mathcal{L}$ through Q. Visualize line $\mathcal{L}$ as hinged at Q. The arbitrariness of line $\mathcal{L}$ will give a multiplicity of choices for point F. In Figures 7.11b and c, dashed lines and hollow dots indicated lines and points that depend on the arbitrary choice of line $\mathcal{L}$. Denote the intersections of $\mathcal{L}$ with lines CD and CB as P and Q, respectively.

The final step (Figure 7.11c) is to draw two lines, through AP and ER. Point F is the intersection of these two lines. Then ABCDEF is a mystic hexagram because pairs

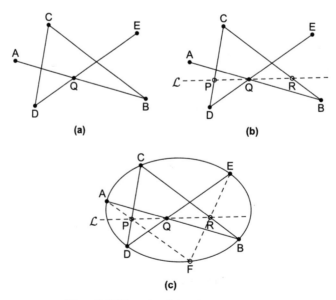

Figure 7.11 Five points determine a conic section.

of opposite sides intersect in the collinear points P, Q, and R. From the converse of Pascal's theorem, the points ABCDEF must lie on a conic section.

The results of this section have the following astronomical applications. If you believe (mistakenly) that a planetary orbit is circular, then knowledge of three positions of the planet is sufficient to determine the center and radius of the circle. On the other hand, if you believe (correctly) that the orbit is elliptical, then knowledge of five positions is needed to completely determine the ellipse.[3]

This chapter gives a glimpse of the rich mathematics of the ellipse and includes results that will be needed later. As a model for planetary orbits, the ellipse replaced the awkward circular makeshifts of the ancient astronomers. The next chapter is the climax of our story—how Kepler removed the circle from the central role prescribed by Plato and pointed the way for Newton to discover the law of universal gravitation.

8

THE WAR WITH MARS

For Mars alone enables us to penetrate the secrets of astronomy which
otherwise would remain forever hidden from us.
Johannes Kepler, *Astronomia nova* (1609)

From antiquity, the fixed stars have been observed in lockstep circular motion about
the celestial poles. However, day-to-day charting of the movements of Mars and the
other planets demonstrated violations of this pattern—yet in a manner that suggested
underlying regularity.

Taunting the ancient Greek astronomers, the planets clothed the complexity of their
motion with specious simplicity. The astronomers took the bait and devised elegant
theories. The planets mocked the astronomers' efforts by departing only subtly from
their predictions. The true nature of the planets might have remained forever hidden
but for the boldness of Mars. For it was Mars that called attention to himself with his
rash antics—exaggerating movements that the others kept elusively small. *Here I am.
Look at me!*

Mars succeeded in inflicting prickly exceptions that led the ancient astronomers to
amend their theories with makeshift patchwork. Eventually, Mars became the target
of an exhaustive campaign of measurement and calculation. Around 1600, Johannes
Kepler entered the war on Mars. He mounted an attack entailing six years of study and
a thousand pages of calculations, and Mars finally yielded its secrets.

Kepler undertook this task as an assignment from his employer, Tycho Brahe.

8.1. TYCHO BRAHE

Tycho was born in 1546 of noble Danish parents. He entered the University of
Copenhagen at the precocious age of thirteen and later studied at the Universities of
Wittenberg and Rostock in Germany. He became entranced with astronomy, to the
dismay of his family who wanted him to become a lawyer or a diplomat. In 1576,
King Frederik II of Denmark offered Tycho the island of Hven near Hamlet's castle
of Elsinore, where Tycho built an astronomical observatory, which he named Urani-

borg, after Urania, the goddess of the sky. Tycho educated his younger sister Sophia (1556–1643), who assisted him with his astronomical observations. Sophia was, in her own right, a noted astronomer, horticulturist, and historian.

Tycho, a wealthy man, employed a large staff to assist him with his astronomical observations. Figure 8.1 shows Tycho together with some of his assistants. Although Tycho generally used a portable sextant, the most spectacular astronomical instrument at Uraniborg was a huge quadrant built into a precisely oriented north-south wall. This instrument was used to measure the angle at which a star or planet in the southern sky crossed the meridian—the great circle through the north and south celestial poles and local zenith. To complete an observation, a clock was needed to determine not only the angle but also the time of transit. Tycho's mural quadrant could measure angles with an accuracy of one minute of arc. (Sixty minutes equals one degree. The full Moon spans about thirty minutes of arc.) Figure 8.1 shows the mural quadrant surmounted by a painted mural of Tycho and views of activities at Uraniborg. Tycho points to a slit in the ceiling through which observations were made. On the right, we see an observer, perhaps Tycho himself, noting an angle of transit. Below, one assistant reads the clocks and another writes a record. Accurate measurement of the time was an essential element of the observation. Accurate mechanical clocks were not available in Tycho's time, but the movement of the fixed stars provided an accurate standard of time.

Tycho incurred the displeasure of King Christian IV by abusing the tenants who lived on his island, and he eventually left Hven in 1597. Through the patronage of the Holy Roman Emperor Rudolph II, to whom he provided astrological forecasts, he established another observatory in a castle near Prague. There he had the good fortune to hire two talented makers of scientific instruments, Joost Bürgi (1552–1631) and Erasmus Habermel (1538–1606). Bürgi and Habermel built astronomical instruments of the highest quality for Tycho. Bürgi was a talented clock-maker, who invented the minute hand in 1577.

Tycho was not a Copernican. The Tychonic planetary system asserted:

1. Earth is motionless.
2. The Sun orbits about Earth.
3. The planets orbit about the Sun.

Tychonic and Copernican systems differed principally in their choice of a frame of reference. The Tychonic system described essentially the same motion with Earth fixed as did the Copernican with the Sun fixed. The Copernican system had the advantage of greater simplicity, and it was the Copernican system that led to Newton's law of universal gravitation.

Figure 8.1 Tycho Brahe's mural quadrant at Uraniborg. From his work *Astronomiae instauratae mechanica* ("Apparatus for the Foundation of Astronomy"), 1598.

In 1600, Tycho received a copy of an astronomical work called *Mystery of the Cosmos* by a young astronomer named Johannes Kepler. Among other things, this book contained Kepler's theory that the spacing of the planets was explained by a construction involving the five regular polyhedra, the Platonic solids—tetrahedron, cube, octahedron, dodecahedron, and icosahedron. Tycho happened to be looking for a mathematical assistant, and Kepler got the job.

Tycho died in 1601. The unlikely story was told that Tycho died of a burst bladder because, from politeness, he neglected to excuse himself from dinner to relieve himself. However, in 1991, examination of exhumed remnants of his beard suggested that he died of poisoning from mercury ingested a day or so before his death.

After Tycho's death, Kepler was named Imperial Mathematician at Prague by Rudolf II, emperor of the Holy Roman Empire. Although Rudolf was most interested in astrological forecasts, Kepler made discoveries during his tenure that would change astronomy forever.

8.2. KEPLER

Kepler was firmly committed to the Copernican system, which he learned from Michael Maestlin (1550–1631), his teacher at the University of Tübingen. On the other hand, Tycho, contrary to Copernicus, proposed that the Sun orbits Earth and the planets orbit the Sun. It is greatly to Tycho's credit that he recognized Kepler's ability and hired him in 1600 although they disagreed on fundamental questions.

Tycho assigned Kepler the task of interpreting Tycho's twenty years of meticulous observations of the planet Mars. Studying Tycho's mountain of data occupied Kepler for six years and produced hundreds of pages of computations. He called it his "war with Mars."

Kepler brought to this task a belief in heliocentrism based on theology. Kepler had a strong belief that God made the universe according to a mathematical plan. It was Kepler's obsession to understand as much as possible of that plan. He believed that God made the Sun, not only the source of warmth and light, but also of the motive power of the planets. Of course, the full explanation of this motive power would have to wait for Newton's dynamics.

Eventually, Kepler placed the center of the solar system exactly at the Sun[1] instead of following Copernicus who had put it at a point *near* but not coinciding exactly with the Sun. Kepler expressed this idea by saying that his theories, unlike those of Copernicus, were "physical" and not merely "mathematical."

Mars was an excellent choice by Kepler for careful study. Mars is convenient to observe. The inferior planets Venus and Mercury are often lost in the Sun's glare, especially when in retrograde, but the superior planet Mars does not have this problem. Mars exhibits more rapid motion than Jupiter or Saturn. However, Mars has an advantage that goes beyond ease of observation. Mars departs from the behavior predicted by the deferent-epicycle model more strongly than any planet except Mercury.

Kepler used Tycho's data to make immense strides in the understanding of the solar system. It became clear to Kepler that the existing theories of planetary motion, including those of Copernicus, were incompatible with the data. The degree of incompatibility was subtle—a discrepancy of four minutes of arc. Thus, Kepler utilized Tycho's capability to measure angles with an accuracy of ± 1 minute of arc.

Kepler's success was made possible through the following circumstances:

1. Tycho's passion to observe and measure the movement of the planets. He used the most accurate astronomical instruments available and kept exhaustive records over many years.

2. The ability of Tycho's assistants, Joost Bürgi and Erasmus Habermel, to make devices for measuring—even before the invention of the telescope—the position of stars with accuracy of ± 1 arcminute.

3. Kepler's willingness to study the resulting data with dogged persistence. He spent six years studying the data relating to the motion of Mars and finally realized that this data would not fit any existing theory.

4. Finally, Kepler's use of his mathematical knowledge to fit the data to an entirely new theory of planetary motion.

Tycho carried out precise observations of planetary motion. It was the records of these observations that provided Kepler with data to support the existence of *elliptical* orbits. Kepler's three laws of elliptical planetary motion foreshadow Newton's discovery of universal gravitation—that the motion of the planets conform to the same dynamical laws that can be observed at a billiard table or an apple orchard.

Kepler chronicled his war with Mars in 1609 in his book *Astronomia nova*.[2]

Analyzing Tycho's data stimulated a storm of astronomical theories in Kepler's brain. From this maelstrom emerged Kepler's three laws of planetary motion (see Section 8.2), which remain today fundamental principles of planetary astronomy. Kepler stated his three laws—supremely important scientific milestones—without the clarity and emphasis that they deserve: the first two in *Astronomia nova* and, later, the third law in *Harmonice mundi* (1619) ("The Harmony of the World").

Kepler believed that Copernicus's heliocentric solar system was a physical reality and not merely a mathematical contrivance. Kepler may have disagreed with Copernicus on the precise shape of the orbits, but Kepler accepted the strong evidence that the planets travel around the Sun in periodic orbits. For example, he was quite certain that Mars returns to exactly the same point in space in about 687 days. Using the synodic period of Mars to compute this number, the Martian year, was discussed in Section 4.3. Ptolemy knew this number, but to him it was the period of rotation of the center of an epicycle about a deferent—merely a mathematical artifice.

Copernicus claimed that the planets orbit about a point close to, but different from, the Sun. Kepler worked with similar theories, but, after much struggle, he showed that the Sun was the sole shepherd of the solar system—that the Sun could keep the planets in their orbits without any phantom influences.

The Vicarious Hypothesis

Kepler's *Astronomia nova*, describing his war with Mars, discussed his first model for the orbit of Mars, which he called the *vicarious hypothesis*. He used the term "vicarious" in retrospect. In recounting his research, he believed that this hypothesis was vicarious because it was a stand-in for the true hypothesis that he discovered at a later time. The vicarious hypothesis is illustrated in Figure 8.2. This model was a circle—the equant of Ptolemy discussed in Section 4.5—with the following changes:

- Earth E in Figure 4.10 is replaced by the Sun S in Figure 8.2.
- The circle represents the orbit of Mars instead of a deferent circle.
- The distances $\overline{QC}$ and $\overline{CS}$ are no longer assumed equal. In Kepler's model for the orbit of Mars, the ratio of these distances was determined empirically.
- Kepler's vicarious hypothesis, in agreement with the Ptolemaic equant in Figure 4.10, assumes that the center C of the equant circle lies on the line QS connecting the Sun S and the equant point Q.

In Figure 8.2, according to the vicarious hypothesis, Mars M travels on a circular orbit centered at C, and its angular velocity with respect to the equant point Q is constant. Kepler gave himself the daunting task of using this model to calculate the orbit of Mars based on Tycho's data. Kepler wished to use Tycho's Mars data in order to find the correct placement of the equant point Q and the Sun S in Figure 8.2. His first difficulty was that Tycho's observations were made from Earth, but Kepler needed sightings of the Mars from the *Sun* instead of from Earth. This problem was solved by using Tycho's sightings of Mars at *opposition*—a time when sightings of Mars from Earth coincide with sightings from the Sun (see Section 4.1). Kepler used Tycho's data for Mars oppositions observed in 1587, 1591, 1593, and 1595.

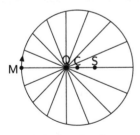

Figure 8.2 The *vicarious hypothesis* asserts that Mars M moves about the Sun S on a circle centered at C with constant angular velocity with respect to the equant point Q. The intersection points of the line segments from the equant point Q to the circumference of the circle are positions of Mars at sixteen equally spaced time intervals. Point M shows Mars at aphelion, the point where Mars is farthest from the Sun.

Kepler used two types of data: *true anomaly* and *mean anomaly*. True anomaly consists of the ecliptic longitude at each of these oppositions— the data that Tycho recorded from his observations of Mars. Mean anomaly is the time of the observation scaled as the longitude at which Mars would have been observed if its motion were uniform. In the equant model, illustrated in Figure 8.2, the mean anomaly of an observation is the longitude of Mars observed from the equant point Q—the point from which the angular velocity of Mars is assumed to be uniform.

Figure 8.3 illustrates the difficult problem that Kepler solved to find the equant orbit that was the best fit for the four chosen oppositions of Mars. This figure, based on a diagram in Kepler's *Astronomia Nova*, illustrates Kepler's method for using the data of the four oppositions. Figure 8.3a shows the longitudinal directions SM_1, SM_2, SM_3, and SM_4, the true anomalies, of Mars at each of the four oppositions, and Figure 8.3a shows the corresponding mean anomalies. Kepler devised a method of successive approximations to achieve the result shown in Figure 8.3c in which Figure 8.3a is superimposed on Figure 8.3b using translation only (no rotation allowed) such that the intersections of the rays representing the true and mean anomalies lie on a circle centered at C.

The problem can be visualized graphically as follows:

- Make a copy of Figure 8.3a on a transparent sheet of plastic.
- Place this sheet on top for Figure 8.3b.
- Move this sheet by translation only (preserving the horizontal and vertical directions) until the intersections of SM_1 with QM_1, SM_2 with QM_2, SM_3 with QM_3, and SM_4 with QM_4 become four points on a circle with center on the line SQ, as shown in Figure 8.3c.

If this difficult task is achieved, then the circle centered at C in Figure 8.3c is a model for the orbit of Mars according to the vicarious hypothesis, with the Sun at S and the equant point at Q.

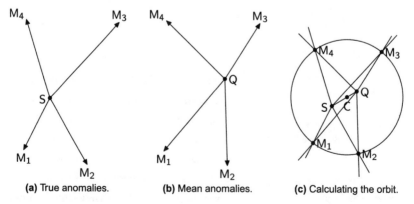

(a) True anomalies.   **(b)** Mean anomalies.   **(c)** Calculating the orbit.

Figure 8.3 The vicarious hypothesis. Kepler used four of Tycho's observations of Mars at opposition to estimate its circular orbit centered at C together with the positions of the Sun S and the equant point Q.

Kepler found a method of successive approximations for solving this problem. Kepler's solution involved laborious computations, but he found an acceptable solution after making seventy iterations of his numerical procedure.

Note that Kepler needed data from at least *four* Mars oppositions. Three data points are not sufficient because *any* three noncollinear points determine a circle.

Now Kepler was able to compare Tycho's other data of observations of Mars with the positions predicted by the equant model. Kepler found that the equant model could be used to make predictions of *longitude* that were accurate within one minute of arc, the width a pencil lead seen at a distance of eight feet—an error that was consistent with the accuracy of Tycho's observations.

Nevertheless, Kepler was dissatisfied with this result because this version of the equant model accounts for longitude only. Kepler believed that the model could have physical meaning only if it gave accurate values for both longitude and latitude. Kepler held this position despite the fact that astronomers before him, from Ptolemy to Copernicus, were all content to use separate mathematical models for longitude and latitude.

Kepler proceeded to deal with latitude by assuming, correctly, that the orbit of Mars lies in a plane containing the Sun making a small angle with a corresponding plane for the orbit of the Earth, the plane of the ecliptic. This angle, the angle of inclination of the orbit of Mars, is now known to be $1.85061°$. Kepler found that the best fit of the equant model with an inclined orbit of Mars showed a disagreement from certain of Tycho's data by eight minutes of arc. But Kepler knew that Tycho's data could have an error of no more that one minute of arc. This tiny disagreement by eight minutes of arc was sufficient to lead Kepler to reject the equant model—that is, to reject the vicarious hypothesis.

Kepler now proposed that the orbit was an oval instead of a circle. He finally settled on an ellipse—a choice that turned out to be marvelously accurate.

Kepler's Three Laws

...Kepler knew ye Orb to be not circular but oval & guest it to be Elliptical...
Isaac Newton, *Letter to Edmund Halley, June 20, 1686.*

In the above quotation, Newton seems to downgrade Kepler's assertion that orbits are elliptical to a mere guess. One might think that Kepler settled on the elliptical orbit because it was an oval curve with the next level of complexity beyond the circle, and it fit the data adequately. But at a certain point Kepler realized that the ellipse fit planetary motion in a further remarkable fashion shown in Figure 8.4. If the orbit of Mars is an ellipse with the Sun at a focus, then the data confirmed that the line from the Sun to Mars sweeps out equal areas in equal times. Thus Kepler's first and second laws were conceived.

I. The orbit of a planet is an ellipse with the Sun located at a focus.

II. A line segment from the Sun to the planet sweeps out equal areas in equal times.

At first glance, Kepler's first law might appear to be just another scheme of curve fitting such as was practiced by Kepler's predecessors, but his second law takes Kepler's total assertion—the first two laws as a unit—to a new higher level of physical meaning. Kepler's second law does not just double the strength of the first law—it enhances it by an order of magnitude. As we will see in Section 10.2, Newton would show that Kepler's second law is a special case of an important principle of dynamics, the *conservation of angular momentum*.

It is inconceivable that Kepler could have formulated these laws without liberally experimenting with geometrical diagrams. His book *Astronomia nova* has a great many complex geometrical figures. These diagrams are used mainly to fit Tycho's data to the correct orbital shape.

Kepler's three laws of planetary motion elaborate his discovery of elliptical orbits. They were called "laws," not by Kepler himself, but by later astronomers after extensive observational confirmation. Kepler's laws remain today the basis of the positional astronomy of the solar system. Kepler's laws led Newton to his Law of Universal Gravitation. Newton put Kepler's three laws in a setting that seemed to explain the entire universe. Kepler discovered the first two laws toward the end of writing *Astronomia nova* (1609), and the third law, to be discussed below, appears only in the later work *Harmonice mundi* (1619). Kepler's three laws mark the end of the centuries-old effort to explain the motion of the planets using ad hoc compositions of uniform circular motions.

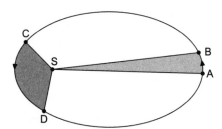

Figure 8.4 Kepler's first law: A planet travels in an elliptical orbit with the Sun S at a focus.

Kepler's second law: The planet travels from A to B and from C to D in equal times if, and only if, the areas of the darkly and lightly shaded regions are equal.

Logarithms

For his war with Mars (1600–1605), Kepler did all of his arithmetic calculations laboriously by hand. This effort would have been greatly eased if Kepler had used logarithms for his calculations—as can be appreciated by those of us who finished our mathematical schooling before 1970, the year when hand calculators replaced logarithmic tables and slide rules.[3]

Logarithms were being invented during Kepler's war with Mars—simultaneously by the Scottish mathematician John Napier (1550–1617), and, ironically, by Kepler's colleague at the Prague observatory, Joost Bürgi. Napier published his tables in 1614 and Bürgi in 1620. Henry Briggs (1561–1630), assisted by Napier, published in 1617 a table of logarithms to the base 10. Briggs's tables won acceptance as a standard calculating tool.

Kepler learned, with great enthusiasm, about logarithms from Napier's publication, not from Kepler's coworker Bürgi. "A prophet is not without honor, save in his own country, and in his own house."[4] When Kepler mastered logarithms, he used them in his computations. In this he was chided by his mentor, Michael Maestlin, for taking unseemly shortcuts. In 1624, Kepler published his own eight-figure table of logarithms. Kepler used logarithms extensively in the computation of his *Rudolphine Tables* (1627), the most accurate astronomical tables in existence at the time of their publication.

Kepler's third law

On March 8, 1618, Kepler discovered his third law. This date, the precise moment of his insight, is documented in *Harmonice mundi*. Kepler did not run naked through the streets crying "Eureka"—as Archimedes did on discovering the law of buoyancy. A sudden insight comes to all of us at times. In Kepler's case, his insight came after intense study and experimentation with the data. Here is a statement of Kepler's third law of planetary motion.

III. The ratio of the squares of the periods of any two planets is equal to the ratio of the cubes of their average distances from the Sun.

Kepler stated his third law in this form, but he was guilty of some imprecision. It is not meaningful to speak of average distance from the Sun without specifying the average with respect to distance, time, or some other variable. For an elliptical orbit, it is possible to define the average in various ways to obtain various results. In fairness to Kepler, it should be said that for nearly circular elliptical orbits these averages are close together and that this kind of quibble was beyond the mathematics of Kepler's time.

Defining this average in a certain rather complicated way, one obtains the semimajor axis (see Figure 7.1b) of the elliptical orbit of the planet. Replacing "average distance" with "semimajor axis" gives a meaningful form of Kepler's third law. It should be added that, in this form, Kepler's third law can be verified with precision by modern instruments.

III'. The ratio of the squares of the periods of any two planets is equal to the ratio of the cubes of the semimajor axes of their elliptical orbits about the Sun.

The planetary data that relate to Kepler's third law, the periods and average distances from the Sun, were known to Copernicus and even to Ptolemy who knew the data as rotation rates and radii of deferents and epicycles.

Kepler's third law can be seen by examining logarithms of astronomical data. Kepler's third law becomes obvious if one makes a double logarithmic plot of the periods of planets (in years) against their average distances from the Sun (in astronomical units), as shown in Figure 8.5a. Kepler did not have this graphical tool at his disposal. Indeed, our widespread use of graphs today to analyze numerical data is largely a twentieth-century innovation. The absence of these tools made the problem more difficult, but not impossible.[5]

In Figure 8.5a, a straight line connects the data points perfectly. In a double logarithmic plot, a straight line indicates a *power law*—a relationship between the horizontal variable x and the vertical variable y of the form $y = kx^r$ for suitable constants k and r. In other words, all points (x, y) on such a line must satisfy the equation $y = kx^r$. Thus a double logarithmic plot is an ideal test to determine if data tend to satisfy a power law. In Figure 8.5b, a few lines are shown together with the formulas that generate them, and the heavy line is identical with the diagonal line in Figure 8.5a, giving the relationship between planetary periods and distances. Denoting R the mean distance from the Sun in AU and T the period in years, Figure 8.5a indicates the relationship $R = T^{2/3}$, which is equivalent to Kepler's third law. The constant of proportionality happens to be one because of the choice of units (AU and years).

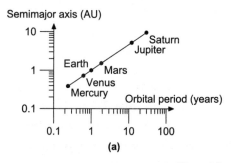 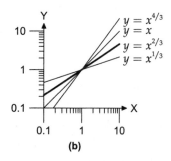

Figure 8.5

Kepler's third law was obtained empirically, searching the data for possible patterns. Kepler saw a musical meaning in this law—that the relationship between the mean distance and period of planet was analogous to the relationship between the length of a vibrating string and its pitch.[6]

Kepler's third law is the fruit of many years of study of this problem. Today, we see his third law as a correct and brilliant solution of an extremely difficult problem.

As we will see, Newton showed that his law of universal gravitation provides a beautiful explanation of Kepler's three laws.

It is interesting to compare Kepler's third law with an empirically derived explanation of the spacing of planetary distances known as *Bode's law*, a theory that is still waiting to be justified or debunked.

Bode's law

What is now called Bode's law was actually discovered in 1766 by Johann Daniel Titius and later published, without attribution, in 1772 by Johann Elert Bode. Thus, Bode's law should be named Titius's law.

Bode's law asserts that the average distances of the planets from the Sun is approximately equal to $0.4 + 0.3 \times 2^k$ AU, where k takes the values $-\infty, 0, 1, 2, \dots$." The planets up to Uranus fit Bode's law within 5%. (Neptune is off by about 25%.) When Bode's law was first stated, neither the asteroid belt nor the planets beyond Saturn were known. Initially, the rank $k = 3$ had to be skipped, and Jupiter was assigned the number 4. Remarkably, the asteroid Ceres was discovered and filled the $k = 3$ gap nicely with an error of about 1%. The planet Uranus was discovered, and its distance from the Sun was within 2% of the prediction of Bode's law. Bode's law remains in a scientific limbo awaiting final judgment.

Kepler's three laws can be verified with much greater accuracy than Bode's law, but this is not the reason why Kepler's laws are considered a towering discovery and Bode's law of minor importance. When Kepler's laws were discovered they had the same defect as Bode's law today: they had no explanation based on fundamental principles of physics. It was Newton who raised Kepler's laws to the high pedestal where they rest today.

Kepler's three laws mark the beginning of astronomy as a *modern* physical science. Before Kepler, each planet required a specially tailored mathematical theory—an equant here, an eccentric there, a Tusi couple here, an extra epicycle there. Modern science looks for greater generality. Today, physical science has a different method of accommodating individual variation. Now the standard paradigm for mathematical physical science has two parts:

1. **A general relationship** such as Kepler's three laws of planetary motion. Today, this relationship is often in the form of one or more *differential equations*—equations involving quantities and their *rates*. Newton found such a generalization of Kepler's three laws that governs not only the descent of a falling apple and the motion of the planets, but even the dynamics of the entire cosmos.

2. **Special instances** are specified by *initial* or *boundary conditions*. For example, an initial condition for the motion of Mars could be the position and velocity of that planet at a particular moment.[7] For a vibrating string, a boundary condition could be that the string is fixed at its two endpoints.

Kepler's parallax method

To bring his war with Mars to a successful conclusion, Kepler needed Tycho's accurate data, and he also needed ingenious methods of calculation. Kepler's writings contain a description of one particularly important tool. This technique is independent of any of the previous theories concerning planetary motion, save one—the periodicity of planetary motion. Earth orbits the Sun and returns to exactly the same point in a year's time, and Mars does the same every 687 days.

Kepler's knowledge of the 687-day Martian year gave him a secure foothold in enemy territory. He found a technique for calculating the three-dimensional position of a heavenly body from two-dimensional observation. This tool alone could have provided a sufficient basis for Kepler's major discoveries—in particular, his three laws of planetary motion.

The position of a heavenly body at a moment in time is specified by a pair of angles—for example, ecliptic longitude and latitude, discussed in Section 2.2. Thus, astronomical observations are two-dimensional. All our visual perception is similarly two-dimensional. In our daily lives, we are able to use visual perception to understand the three-dimensional world (1) by sensing movement of objects and (2) by parallax using binocular vision (see Section 2.1). Sensing movement of celestial objects, other than an occasional meteor, was not possible because of their enormous distance. Parallax is a possible astronomical tool, but in Kepler's time, parallax required the measurement of angles too small to be observed with the available astronomical instruments. Kepler began his study of Mars by solving this seemingly hopeless problem.

Kepler wanted to use parallax to deduce the changing distance to Mars as it moves in orbit. He used the movement of Earth in its orbit around the Sun to observe Mars from two positions that are at a great distance from each other. The difficulty is that he needed to juggle with *two* moving planets. As Earth moves to a new position in its orbit, Mars also moves. If there was only some way to make Mars stand still! As always with the juggling art, timing is everything.

Kepler solved this difficulty by using Tycho's observations at multiples of the 687-day Martian year. The advantage is that at each of these observations Mars is in exactly the same position. In antiquity, long before Kepler, it was known that the length of the terrestrial year was $365\frac{1}{4}$ days. Copernicus knew that a Martian year, the length of time for Mars to orbit the Sun, is about 687 days.

Thus, if we observe Mars at an interval of one Martian year, Mars has returned to its original position—it has "stood still." This is shown in Figure 8.6. In 687 days, Mars M returns to its initial point. In the meantime, Earth has traveled almost twice around the Sun—from its initial position E_1 to its final position E_2—about 43° short of two full rotations.[8] *Assuming that Earth's orbit is a circle with the Sun at its center,* a computation finds the distance d between E_1 and E_2 to be about 0.739 AU. (Recall that an astronomical unit [AU] is the mean distance between the Sun and Earth.)

Now parallax can be achieved. Kepler used trigonometry to compute the three-dimensional coordinates of Mars at point M. This method succeeded, in part, because the angular differences between pairs of sightings were measured with sufficient accuracy.

This seemed to be a clever way to use the hundreds of observations that Tycho had accumulated over twenty years. However, there was a snag. Although Earth's orbit is *nearly* a circle with the Sun as center, this approximation is not close enough to give Kepler a sufficiently accurate tracking of the orbit of Mars by the above method. Before proceeding with the charting of Mars's orbit, Kepler must first find a more accurate representation of the orbit of Earth.

The need for correction of a sundial using an analemma (see Section 2.3) stems from the fact that Earth's orbit is more complicated than just uniform circular

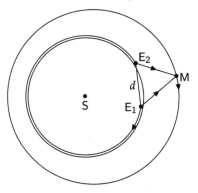

Figure 8.6 Kepler's method to find the three-dimensional position of Mars by triangulation. The inner and outer curves are the orbits of Earth and Mars, respectively. E_1 and E_2 denote the positions of Earth at the start and finish, respectively, of a Martian year of 687 Earth-days—somewhat less than two terrestrial years. The position of Mars M is exactly the same at the start and the end of this period. If the points E_1 and E_2 and the distance d between them are known, then the angular sightings of Mars (right ascension and declination) along $E_1 M$ and $E_2 M$ can be used to calculate by triangulation the three-dimensional position of point M.

motion centered at the Sun. In antiquity, Hipparchus suggested the eccentric model, which improves accuracy of the circular orbit by moving the Sun slightly away from the center of the circle. Further improvement is possible using the equant model (see Figure 4.10).

Surprisingly, Kepler found that observations of Mars can be used to improve knowledge of Earth's orbit.

Help from Mars

Figure 8.7a shows how Kepler used observations of Mars to find an improved estimate for the orbit of Earth. In the following calculation, Kepler assumes Earth's orbit is circular without placing the Sun at the center of the circle. This permits a good fit. Of course, assuming an elliptical orbit would give a still better fit. As in Figure 8.6, M and E_1 are the initial positions of Mars and Earth. After one sidereal Martian year of 687 days, Mars returns to point M and Earth rotates almost two years to E_2. Furthermore, E_3 is the position of Earth after two Martian years: $2 \times 687 = 1,374$ days.

Kepler makes use of Tycho's observations of the ecliptic longitude and latitude of Mars on these three occasions—the directions of each of the line segments E_1M, E_2M, and E_3M. The directions of E_1S, E_2S, and E_3S are the known from the precise position of the Sun in the zodiac—the ecliptic longitude—on each of these three dates. It is important that the above observations do not depend on any speculative theories

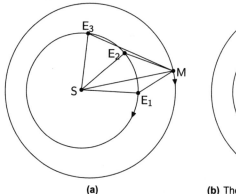

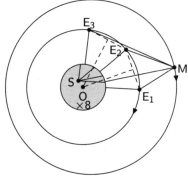

(a)

(b) The eight-fold enlargement shows the center O of the putative circular orbit of Earth is close to, but not identical with, the Sun S.

Figure 8.7 Using Mars to track Earth's orbit. (a) Mars occupies point M initially, at the end of a Martian year of 687 days, and at the end of two Martian years ($687 \times 2 = 1,374$ days). Earth is at points E_1 initially, E_2 at one Martian year, and E_3 at two Martian years. The slopes of lines E_1M, E_2M, and E_3M are known from astronomical observation, and the slopes of E_1S, E_2S, and E_3S are known from the positions of the Sun in the zodiac at each of the three observations. A trigonometric calculation now makes it possible to compute lengths $\overline{E_1S}$, $\overline{E_2S}$, and $\overline{E_3S}$, each as a fraction of the length $\overline{MS}$. Now it is possible to compute the three-dimensional coordinates of E_1, E_2, and E_3. (b) Shows that the three points, E_1, E_2, and E_3 can be fit on a circle, the putative orbit of Earth.

concerning the nature of Earth's orbit. These measurements only make use of the firmly established result that Mars returns to exactly the same position, relative to the Sun, every 687 days.

The lengths $\overline{E_1 S}$, $\overline{E_2 S}$, and $\overline{E_3 S}$ can now be calculated. Each is a different fraction of the distance $\overline{SM}$, which can be computed trigonometrically.

Initially, Kepler believed that Earth's orbit was a circle with the center close to—but not identical to—the Sun. Three noncollinear points such as E_1, E_2, and E_3 determine a unique circle. In Figure 8.7b, the center of this circle is the intersection O of the perpendicular bisectors of the line segments E_1, E_2, and $E_2 E_3$. The darkened circular region in Figure 8.7b is an eight-fold magnification of a detail showing that the intersection point O is close to but different from the Sun S.

This representation of Earth's orbit was inaccurate. Later, Kepler would discover that the orbit is not circular but elliptical. But Earth's orbit is nearly circular, and the errors were small enough to give Kepler a foothold—a good-enough base to track the orbit of Mars using the method shown in Figure 8.6.

Using the parallax method described above, Kepler was able to track the three-dimensional movement of the planets better than anyone before him. He had the ability to track a large number of positions of Mars. Examining four or more such points, he was able to see that, within experimental error, the orbit of Mars lies on a plane containing the Sun with a slight inclination (1.85°) to the ecliptic plane. (Similarly, Earth's orbit lies on a plane containing the Sun—the ecliptic plane. This is the heliocentric interpretation of the fact, known from antiquity, that the ecliptic circle is a great circle on the celestial sphere.)

If Kepler suspected that the orbit of Mars was elliptical, he needed three-dimensional data for least *five* positions of the planet. This follows from Proposition 7.6, which asserts that five distinct coplanar points determine a unique conic section—a consequence of Pascal's Mystic Hexagram. Unfortunately, even Tycho's extensive collection of data contained only a small number of observations of Mars at intervals of 687 days. Furthermore, it is highly unlikely that Kepler knew of Pascal's Mystic Hexagram because Kepler died in 1630, nine years before Pascal made his discovery. Nevertheless, Proposition 7.6 shows that at least five positions of Mars must be known before it is possible to specify, by whatever means, an elliptical orbit.

Part III now tells how planetary astronomy, with Newton's help, became a full-fledged modern science—sufficiently reliable to support the voyages of astronauts.

PART THREE

Enlightenment

9

THE BIRTH OF MECHANICS

Give me a place to stand and I will move the earth.

Attributed to Archimedes *by Pappus of Alexandria, 340?* CE

Ancient astronomers thought that the stars, Sun, Moon, and planets were subject to different rules than ordinary earthly objects. Eventually, Newton showed that the laws of force and motion are the same for a falling apple as for the stars in the sky. Aristotle speculated on the physical laws governing ordinary objects, but Archimedes was the first to treat such matters scientifically.

Newton wove together two threads of thought—astronomy and mechanics—to create his *universal law of gravitation*. To prepare for the discussion in the next chapter about Newton's contribution, the present chapter deals with the beginnings of the science of mechanics. *Mechanics* is the part of physics that deals with *force* and *motion*. *Statics* is the part of mechanics dealing with forces on objects at rest. *Kinematics* is the mechanics of motion, and *dynamics* is the study of the interaction of force and motion.

In antiquity, mechanics was exclusively statics. The force that was commonly measured was the weight of an object. The equal-arm balance was used for this purpose as early as 5000 BCE. Weight measurement has always been an important activity in commerce. On the other hand, *motion* was not measured accurately until about 400 years ago, when Galileo charted the motion of balls rolling on an inclined plane.

9.1. ARCHIMEDES

The Greek mathematician Archimedes lived in Syracuse, a Greek colony in Sicily. He introduced several mechanical concepts— the center of gravity, buoyancy, mechanical advantage (levers and pulleys), and the hydraulic screw. He also used the lever to compute areas and volumes. For example, by balancing a combination of sphere, cylinder, and cone, he was able to find a formula for the volume of a sphere.[1]

Archimedes is said to have used his discoveries in mechanics to design weapons of war for the defense of Syracuse against the Romans during the Second Punic War—the war in which Hannibal and his army of Carthaginians crossed the Alps to attack Rome. In particular, the Claw of Archimedes is said to have been some kind of crane and grappling hook used to attack hostile ships. Archimedes was killed by a Roman soldier during the invasion of Syracuse.

The following sections discuss three of Archimedes' contributions to mechanics—the concepts of center of mass, buoyancy, and the lever.

Center of mass

The *center of mass* of an object, also known as its *center of gravity*, is its point of balance. For example, a ruler balances at its midpoint. The centers of mass of geometrical objects, absent from Euclid's *Elements*, is first discussed in Archimedes' treatise, *On the Equilibrium of Planes*. Archimedes gives geometrical rigor to center of mass by formulating a list of axioms from which he proved a number of propositions.

Some textbooks begin by discussing the center of mass of a finite collection of mass particles. For example, it can be shown that the center of mass of three particles of equal mass placed at the vertices of a triangle lies on each of the three medians of the triangle. (A median of a triangle is a line joining a vertex with the midpoint of the opposite side.) It follows that the three medians are concurrent.

The center of mass of a two- or three-dimensional geometrical object of constant density is called its *centroid*. A *lamina* is a plane object of constant density—an object that can be approximated by cutting a flat piece of sheet metal. In his treatise, Archimedes considers the centroids of, for example, laminar parallelograms and the triangles. Today, these centroids would be defined using calculus. This is one of many instances in which Archimedes foreshadows the calculus.

Archimedes proves that the centroid of a laminar triangle of uniform density is located at the intersection of the medians of the triangle—remarkably, the same as the center of mass of a triangle with equal point masses at its vertices, as discussed in the previous paragraph.[2] Archimedes' proof is given below in detail because it is an elegant illustration of his mathematical genius and, by any standard, a beautiful example of the indirect method of proof.

One of Archimedes' axioms and two of his propositions are stated here in preparation for this proof. These plausible ancillary propositions are stated here without proof.

The following axiom asserts that proportional magnification (or reduction) of a figure affects the center of mass in the same manner as the figure as a whole.

Axiom 6. The centers of mass of unequal but similar figures are similarly placed.

Proposition 4. If two figures of equal mass do not have the same center of mass, the center of mass of the figure made up of the two figures is the point situated at the middle of the line that joins their centers of mass.

Proposition 10. The centroid of a laminar parallelogram is the point of intersection of its diagonals.

Theorem 9.1. *The centroid of a laminar triangle lies on the straight line joining any vertex to the midpoint of the opposite side.*

Consequently, the three medians of the triangle intersect in a point, and this point is the center of mass.

The following is based on the second of Archimedes' two proofs of this theorem.

Proof. Referring to Figure 9.1, suppose that P is the center of mass of triangle △ABC and that BF is one of its medians. It will be shown that, contrary to what is shown in Figure 9.1, P must lie on BF. The idea of the proof is to decompose the triangle △ABC into three pieces—the parallelogram $\mathcal{P}$ and the two small triangles $\mathcal{L}'$ and $\mathcal{L}''$.

Points D, E, F, H, and J denote midpoints of the segments, respectively, AB, BC, CA, CF, and AF. Triangles $\mathcal{L}'$ and $\mathcal{L}''$ are similar to the large triangle △ABC with a scaling factor equal to 1/2. Line segments EH and DJ are the medians of the small triangles that correspond to the median BF of the large triangle. As shown in Figure 9.1, let Q and R be the centers of mass of the smaller triangles.

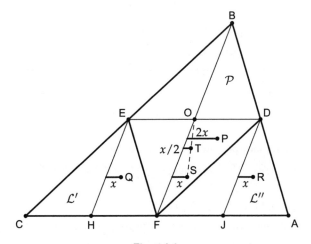

Figure 9.1

By Axiom 6, the centers of mass of these three triangles must be similarly placed. Since the scaling factor is $1/2$ from the large triangle to the small ones, the horizontal distance from center of mass to median of the large triangle must be twice the horizontal distances from center of mass to median for each of the two smaller triangles. Putting $2x$ as the horizontal distance from center of mass P to median BF, the corresponding distances for the smaller triangles—Q to EH and R to DJ—must be equal to x.

Consider the figure, denoted $\mathcal{S}$, consisting of the union of the triangles

$$\mathcal{L}' = \triangle CEF \qquad \text{and} \qquad \mathcal{L}'' = \triangle ADF$$

By Proposition 4, the center of mass of $\mathcal{S}$ is located midway between Q and R, the point denoted S in Figure 9.1. The horizontal distance between S and the line segment BF is x.

By Proposition 10, the center of mass of parallelogram $\mathcal{P}$ is the intersection O of its diagonals.

Because the area of parallelogram $\mathcal{P}$ is equal to the area of $\mathcal{S}$ (the union of the two triangles), Proposition 4 asserts that the center of mass of the union of $\mathcal{P}$ and $\mathcal{S}$—in other words, the entire triangle $\triangle ABC$—is the point T midway between O and S. Furthermore, the horizontal distance to line BF is equal to $x/2$. Point T, since it is the center of mass of the entire triangle $\triangle ABC$, should, unlike Figure 9.1, coincide with P, which was assumed to have horizontal distance to line BF equal to $2x$. If these two points coincide, then $x/2$ must be equal to $2x$. But this is possible only if $x = 0$—that is, only if the center of mass P of the entire triangle $\triangle ABC$ is located on the median BF. $\qquad\square$

Buoyancy

It was on discovering the principle of buoyancy that, according to legend, Archimedes ran naked through the streets crying "Eureka." Two works of Archimedes dealt with buoyancy—*On Floating Bodies, Books I and II*. Book I deals with basic matters, and Book II makes specific computations for a floating paraboloid. The principle of buoyancy, also known as Archimedes' principle, is the following:

Principle 9.1. *Buoyancy. Suppose that an object is immersed in a motionless fluid. Then the object is subject to an upward buoyant force equal to the weight of the fluid displaced by the object.*

Archimedes proved Principle 9.1 by using assumptions concerning the shape of the submerged object. Generality concerning shapes is a modern concept that did not exist in antiquity. I prefer to establish Principle 9.1 with the following line of thought.

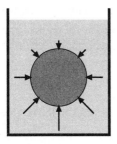

Figure 9.2 Pressure on the immersed object varies with depth.

The buoyant force is the resultant of the pressure at all points on the surface of the object. This force depends only on the shape of the object and has nothing to do with the interior of the object. Imagine an invisible twin of the object. The twin has the same shape as the object, but its interior is empty. Further imagine that the fluid flows through the twin's surface without obstruction because the twin is merely an imaginary surface. Remove the object and put the twin in its place at the same depth. The buoyant force on the twin is exactly the same as the buoyant force was on the original object. The fluid was assumed motionless and, therefore, so is the twin. Because the twin is motionless, the buoyant force exactly cancels the weight of the twin—the weight of the fluid displaced by the object—as claimed by Principle 9.1.

It is interesting that the above argument makes no assumption how pressure interacts with the surface of the object or how smooth or rough its surface might be. The above argument applies to the spherical object shown in Figure 9.2 or to a rough object like a submerged sponge. There is no need to discuss how the force of gravity creates the pressure in the fluid or how this force permits a fluid to be motionless.

Levers and pulleys

Archimedes was quite aware of the practical uses of the levers and pulleys to lift heavy objects. He used this knowledge to design war machines to defend Syracuse against the Romans. The historian Plutarch wrote that Archimedes demonstrated the effectiveness of a system of pulleys by using his personal strength alone to move a fully loaded war ship.

Figure 9.3 illustrates the use of (a) a lever and (b and c) a pulley system to achieve *mechanical advantage*—the ratio of the weight of a object to force needed to lift it. In Figure 9.3, the lever and pulley systems are in equilibrium although weight #2 is twice as heavy as weight #1—illustrating a mechanical advantage of 2 : 1.

In Figure 9.3a, d_1 and d_2 are the distances from the fulcrum to the respective centers of mass of the weights #1 and #2. One of Archimedes' propositions in *On the Equilibrium of Planes* implies that the lever shown in (a) is in equilibrium if the ratio $d_1 : d_2$ is equal to 2 : 1, the ratio of weight #2 to weight #1.

Block and tackle is the nautical term for arrangements of pulleys and ropes similar to the one shown in Figures 9.3b and c.[3] The system, shown in (b) and (c), is in equilibrium if, as before, weight #2 is twice as heavy as weight #1. No doubt Archimedes had a mathematical explanation for this equilibrium because he was well acquainted with pulley systems and used them for practical purposes. However, no writing of Archimedes concerning pulley systems has come down to us. In particular, he gives no general theory that covers both the lever and the pulley systems.

Virtual work

About 2,000 years after Archimedes, the Swiss mathematician Johann Bernoulli (1667–1748) defined *virtual work*, an important general principle of statics that explains the mechanical advantage of both the lever and the pulley systems.

Work has a technical meaning that is based very loosely on the idea of "effort." If the point of application of a force moves, positive or negative work results. Although work is a more general concept, the following definition suffices for the current application in which the only force under consideration is the weight of an object and the only movement is vertical up or down.

Definition 9.1. The work associated with moving a weight w vertically a distance d is defined as $\pm wd$ where the plus or minus sign is used depending on whether the displacement is down or up.

Virtual work is the total work associated with small hypothetical displacements of the weights—taking account of the interdependence of all displacements of the system. For example, in Figure 9.3a, if δ_1 and δ_2 are the respective displacements of weights #1 and #2, then $\delta_1 : \delta_2 = 2 : 1$. If one of the displacements is down, the other must be up. Taking account of the plus and minus signs, the total work $w_1\delta_1 + w_1\delta_2$ is equal to

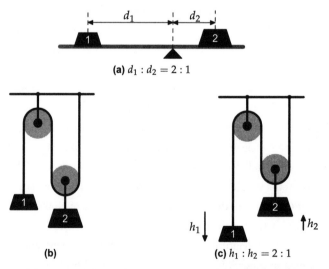

(a) $d_1 : d_2 = 2 : 1$

(b)

(c) $h_1 : h_2 = 2 : 1$

Figure 9.3 Lever (a) and pulleys (b and c) with 2 : 1 mechanical advantage. Weight #2 weighs twice as much as weight #1. In other words, denoting the magnitudes of the weights w_2 and w_1, respectively, the ratio $w_2 : w_1$ is equal to 2 : 1.

zero.[4] This corresponds to the fact that the lever is in equilibrium—an instance of the following general principle.

Principle 9.1.   *Virtual work. A system is in equilibrium if, and only if, for any choice of small virtual displacements consistent with the constraints of the system the total virtual work is zero.*[5]

Figure 9.3c shows the displacements h_1 and h_2 of the pulley system in (b). The movement of the rope is such that the ratio of these displacements $h_1 : h_2$ is equal to 2 : 1. Because the ratio of the weights is $w_1 : w_2 = 1 : 2$, the total work is zero— just as it was for the lever in Figure 9.3a. Thus virtual work gives a common theoretical framework for both the lever and the pulley systems.

The Principle of Virtual Work asserts that the lever and the pulley systems in Figure 9.3 are in equilibrium. Each case illustrates a mechanical advantage of 2 : 1. The practical implication is clear. Removing weight #1, one can pull with a force of 1 pound to balance weight #2 weighing 2 pounds, and pulling with greater force will raise weight #2.

9.2. GALILEO

In Section 6.3, the contributions of Galileo to astronomy were discussed. This section is concerned with Galileo's work in mechanics—largely contained in his work *Two New Sciences*.

The scientific method, as discussed in Section 1.2, consists of an alternation between theory and observation or experiment. In astronomy, observations are feasible but are not experiments. Galileo's experiments in mechanics were probably the first scientific experiments ever performed.

Galileo performed an experiment in which he measured the acceleration of a ball rolling down an inclined plane.[6] He was able to measure the time for the ball to descend specified intervals of distance. Galileo used an inclined plane because a freely falling body fell too rapidly for such measurements. For the measurement of time, Galileo used a water clock. He also arranged objects like frets of a guitar along the inclined plane so that the ball made clicks as it passed each of them. Galileo arranged the frets so that he heard equally spaced rhythmic intervals. This happened when the distance intervals between the frets were arranged proportional to the odd numbers 1, 3, 5,.... From this experiment, Galileo concluded that the distance descended was proportional to the square of the time. That is, if s is the distance descended in time t, then $s = ct^2$, where c is a constant of proportionality.

Galileo considered the motion of projectiles. He understood that projectile motion has two components—a downward vertical component with gravitational acceleration and a horizontal component with constant speed. He was the first to see that these components combine so that the path of a projectile is always a parabola, a curve named and studied by the ancient Greek mathematician Apollonius of Perga (see Section 7.1).

Galileo also discovered by experiment that the *period* of a pendulum, the length of time for one oscillation, is proportional to the square of the length of the pendulum—independent of both the initial displacement and the weight of the bob.

Thus, Galileo anticipated Newton's Laws of motion and universal gravitation.

In his study of mechanics, Galileo did not incur interference from the Catholic Church, as he did in the field of astronomy. However, there were other very serious difficulties.

Galileo's handicaps

Galileo's first handicap was the general acceptance, by the scholastic philosophers and others, of the erroneous ideas of Aristotle concerning motion. For example, Aristotle wrote that motion could only take place with the constant presence of a motive force—in contradiction to the principle later enunciated by Newton that a mass particle with constant velocity continues its linear motion until a force deflects it. Neither Aristotle nor his followers conceived of studying motion, or any other scientific topic, by experiment.

Galileo's second handicap was his lack of experimental tools—in particular, an accurate clock. Galileo had only a primitive method of measuring time—a water clock. He measured a time interval by unplugging and plugging a tank of water and weighing the water released.

Galileo's third handicap was that he lacked mathematical tools. Some of these tools he ignored and others were undiscovered in his time.

Decimal fractions

Galileo did not make use of decimal fractions although they were used in the financial transactions of his time. Simon Stevin (1548–1620), a Belgian, published in 1585 a book concerning decimal fractions, *De Thiende (The Art of Tenths)*. Whether Galileo heard of them or not, decimal fractions were known in his time, but, incredibly, he used only whole numbers in recording his experimental data—a number usage that was inappropriate for recording continuously varying quantities.

Algebra

Algebra was a well-developed discipline in Italy in Galileo's time and yet he made no use of it. For example, Girolamo Cardano (1505–1576) published in 1545 a treatise on algebra, *Ars Magna*, which included, among other things, a method for solving cubic equations.[7] In Cardano's time there were public contests in which mathematicians matched wits in solving equations. Galileo would not have understood the formula

$s = ct^2$ mentioned above in connection with his work with the inclined plane. Although Galileo made no use of algebra, it would have been very useful to him.

Calculus

Calculus would have been extremely useful to Galileo, but calculus did not exist in Galileo's time. Newton and Leibniz are credited with inventing the calculus late in the seventeenth century, too late for Galileo. However, Fermat invented one of the key calculus concepts, the derivative, much earlier.

Before the derivative concept, the *instantaneous velocity* of an object moving at a variable rate was incapable of precise definition, and *acceleration* was even more out of reach. Average velocity over a time interval was easily defined as the distance traveled divided by the length of the time interval. Instantaneous velocity is the average over an "infinitesimal" time interval, but this concept did not achieve clarity until Fermat (1601–1665) studied it, and full mathematical rigor was established only in the mid-nineteenth century. Velocity was fundamental to Galileo's research. Although he struggled with this concept, his writing on mechanics, lacking mathematical rigor, was necessarily somewhat imprecise. As a pioneer in the study of the dynamics of motion, Galileo created a new vocabulary for this new science.

This chapter discussed the early contributions of Archimedes and Galileo to the science of mechanics. Their work prepared the way for Newton, who greatly expanded this science. The next chapter tells how Newton unified mechanics and astronomy with his discovery of the law of universal gravitation.

10

THE ASTRONOMICAL ALCHEMIST

Nature and Nature's laws lay hid in night.
God said, *Let Newton be!* and all was light.
Alexander Pope, *epitaph intended for Sir Isaac Newton (1730)*

Late in the seventeenth century, our understanding of the solar system was transformed by a revolutionary insight—that the motions of the Sun, moon, and planets were governed by the same rules as ordinary physical objects. This is the thought that inspired the young Isaac Newton when he saw an apple fall from a tree. Newton made two pivotal discoveries that unified the sciences of mechanics and astronomy: his *laws of motion* and his *law of universal gravitation*.

As illustrated by the above quotation of Alexander Pope, Newton became a celebrity. More than that, he became a deity of the Age of Enlightenment, the eighteenth-century intellectual movement that exalted science and rationality. Even today, there is a desire to ascribe infallibility to Newton's scientific work. However, Newton was a human being, and his masterpiece, *Principia*, contains some minor flaws, as we will see.

It is ironical that Newton was obsessed with alchemy, a pursuit abhorred by the Enlightenment. He kept his alchemical activities and mystical beliefs separate from his scientific work.

Much later, Einstein overturned Newtonian mechanics. Nevertheless, Newtonian mechanics remains the tool of choice for practical applications—mechanical engineering as well as space technology.

Newton was born prematurely into a family of small independent farmers in Woolsthorpe, England. He was not expected to live. Newton's father died before his child was born, Newton's mother remarried, and Newton was raised by his grandmother.

It has been speculated that Newton, together with Albert Einstein and some other notable figures in science and the arts, had Asperger's syndrome, a mild form of autism. It is certain that Newton was eccentric and was a lifelong loner with marginal social skills.

Newton did not distinguish himself particularly in school, but his maternal uncle, William Ayscough, seeing signs of special ability, persuaded Newton's mother to send him to Cambridge where his interest in mathematics was awakened. In 1665–1666, Cambridge was closed because of the plague raging in London, and Newton returned home to Woolsthorpe. During this short time, although lacking a large library and discussions with professors or fellow students, young Newton established the rudiments of scientific discoveries that were sharpened and enhanced in his later life: the binomial theorem, the nature of color, the calculus, and the law of universal gravitation.

10.1. NEWTON'S DYNAMICS

Newton introduced concepts to mechanics that even today are basic to that science. Newton created the principles of dynamics: the part of mechanics concerning forces and motion. He gave precision to the ideas that Galileo was groping for in his experiments measuring the descent of a ball rolling down an inclined plane, mentioned in Section 9.2.

Speed, velocity, and acceleration

Instantaneous speed—the speed with which an object is moving at a particular instant—seems a more self-evident concept today than it was in Newton's time. Today we have a mechanical device, the speedometer, that measures instantaneous speed directly. However, this appearance of simplicity is an illusion. Studying the speedometer's mechanism does not lead to an easy understanding of instantaneous speed.

The derivative

Newton clarified the meaning of instantaneous speed of a moving particle by introducing a new mathematical concept: the derivative. He asserted that the instantaneous speed of a particle was the quotient of the infinitesimal distance traveled in an infinitesimal time interval.

For motion described by a mathematical formula, Newton was able to compute a formula describing the instantaneous speed—using techniques that first-semester calculus students spend many hours learning. For example, as noted in Section 9.2, the descent of a ball on Galileo's inclined plane is described by the formula $s = ct^2$, where c is a constant and s and t are, respectively, the distance traveled and the time elapsed, putting $t = 0$ at the initial rest position. In this case, Newton could calculate that the speed was given by the formula $2ct$.

Newton's ideas were somewhat imprecise because there is no such thing as an infinitesimal interval of time. Bishop George Berkeley (1685–1753) called Newton to task, calling an infinitesimal "a ghost of a departed quantity." Instead of answering this criticism, Newton (together with Gottfried Wilhelm Leibniz (1646–1716) who discovered the calculus independently) created a huge self-consistent discipline, the calculus, which remains to this day the fundamental language of mathematical physics. About 150 years later, Richard Dedekind (1831–1916) and others placed the calculus on a rigorous foundation without using the concept of infinitesimals—nonetheless, preserving Newton's achievements.[1]

The *velocity* of a particle is its speed together with its direction of motion. In the case of one-dimensional motion—motion along a fixed straight line with an arbitrarily chosen positive direction—velocity is speed together with a suitable algebraic sign, plus or minus.

Newton defined speed as the derivative of distance with respect to time and acceleration as the derivative of velocity with respect to time. Leibniz devised a notation for the derivative that is still used today. If the distance s traveled by a particle depends on time t, then the speed, the derivative of distance with respect to time, is denoted ds/dt, and the acceleration, the derivative of velocity with respect to time, is denoted d^2s/dt^2. In the above example in which $s = ct^2$, a calculus student is able to compute

$$\frac{ds}{dt} = 2ct \quad \text{and} \quad \frac{d^2s}{dt^2} = 2c$$

However, this book does not presume that the reader has any technical or computational knowledge of calculus. The symbol ds/dt should be taken as an abbreviation for "the instantaneous rate of change of distance with respect to time." The symbol d^2s/dt^2 means "the rate of change of the rate of change of distance with respect to time," which is usually called *acceleration*.

The above dependence of distance s on time t describes adequately the motion of a particle moving on a one-dimensional linear path. However, the abstract understanding of motion in two or three dimensions leads to the more sophisticated concept of *vector*.

Vectors

It is important to make a distinction between quantities that can be specified by a single magnitude, called *scalars*, and those that require both a magnitude and a direction, called *vectors*. Speed is an example of the former and *velocity* the latter. It is a convention of mechanics that velocity means speed together with the direction of motion.

There are two notations commonly used for vectors.

1. **The boldface font.** For example, velocity is often represented as **v**, while the magnitude of the velocity, the speed, may be written v.

2. **The upper arrow.** In Figure 1(a) in Appendix B, the vector with tail at A and head at B is denoted $\overrightarrow{AB}$.

Newton did not use vectors because *vector analysis* gradually unfolded only in the nineteenth century—due in large measure to the writings of the American physicist/mathematician Josiah Willard Gibbs (1839–1903). A few simple ideas about vectors are helpful in explaining Newton's ideas. In this book, only the most basic concepts of this complex subject will be used. In addition to the *definition and notation* discussed above, *equality of vectors, addition and subtraction of vectors*, and *multiplication of vector by a scalar* are discussed in Appendix B.

A vector is often represented graphically as a line segment with an arrow indicating its direction. The notation for the directed line segment, the vector, from point A to point B is $\overrightarrow{AB}$.

Force and mass

Newton saw that *weight* was not an appropriate fundamental concept for mechanics. Weight has two important generalizations: *force* and *mass*.

Force, as a technical physical concept, quantifies any influence that tends to change the velocity of an object. Push or pull are words commonly used to describe force. Some forces like gravitation do not involve physical contact. When two oppositely directed forces of equal magnitude act on a body, they cancel each other and no actual change of velocity occurs.

Mass is the measure of the quantity of matter.

There is a clue to these two concepts suggested by two methods of weight measurement: the spring scale and the double pan balance (Figure 10.1).

The bathroom scale for home use is usually a variant of the spring scale, but the scale with sliding weights in a doctor's office is a variant of the double pan balance.

The spring scale is based on the fact that, within a suitably small range of movement, the increase in the length of a coiled elastic spring is proportional to the *force* applied.

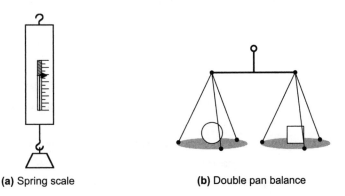

(a) Spring scale   **(b)** Double pan balance

Figure 10.1

This fact is called Hooke's Law after the English scientist Robert Hooke (1635–1703), a contemporary and bitter rival of Isaac Newton. An object on a spring scale reaches an equilibrium in which the applied force is balanced by a force in the opposite direction due to the elongation of the spring. The spring scale can easily be used to measure forces other than weight, but that is not possible for the double pan balance. The equilibrium of the pan balance, based on the principle of the lever (discussed in Section 9.1), is achieved by placing objects of equal *mass* in the two pans. The everyday meaning of weight is mass, the quantity of matter. On the moon, a pound of cabbage is still a pound of cabbage—correctly measured by the pan balance, although the spring balance shows only about one-sixth of a pound due to the weaker lunar gravitation. In mechanics, distinct units of measure are used for mass and force.

Momentum $\mathbf{p}$ is defined as the product of the mass of a body times its velocity: $\mathbf{p} = m\mathbf{v}$. Momentum, a vector quantity, has magnitude and direction (the direction of the motion).

Body or particle

Newton's three laws of motion relate to a body that has mass and moves with a certain velocity. If a body has a single path and velocity, then at any moment in time the body must occupy a single point. By implication, the body is a moving point mass—that is, a *particle*.

The particle concept is an idealization. It is not possible to observe such a thing as a point mass—just as no one has observed a point or line as it is conceived in Euclidean geometry. The motion of a complex object can be studied by considering it as an aggregation of particles. It is a paradox that, although points and particles do not exist as objects in the physical universe, yet they are extremely important and useful abstractions.

Newton's three laws of motion clarified the problem of the dynamics of a particle—including Galileo's experiments with the inclined plane and the pendulum. Newton's Laws are contained in Newton's monumental work, *Principia*, published in 1687 in Latin.[2] The following is a loose translation of Newton's Laws. I have taken the liberty of using the word "particle" instead of "body," the usual English translation of the Latin "corpus."

First Law, the Law of Inertia. A particle at rest remains at rest, and a particle in motion continues to move in a straight line with a constant speed unless an external force acts upon it.

Second Law. The rate of change of momentum of a particle is directly proportional to the impressed force and takes place in the direction in which the force acts.

Third Law. For every applied force there is an equal reactive force in the opposite direction.

The first law describes what is often called the *inertia* of a particle. Here Newton corrects the error of Aristotle who claimed that a body remains at rest unless a force acts upon it. On the contrary, Newton understands that linear motion with a constant velocity describes the motion of a particle in the absence of an external force.

The second law is the most important of the three: the rate of change of the momentum of a particle is equal to the impressed force. In most ordinary circumstances the mass of a particle is constant. Consequently, the rate of change of momentum is proportional to the mass of the particle times its acceleration.

If the mass, acceleration, and the imposed force are m, a, and F, respectively, then Newton's second law states $F = cma$, where the force and acceleration are both in the same direction and c is a suitably chosen constant of proportionality. It is customary to choose units of measurement such that the constant c is equal to 1. In the *cgs* system (centimeter, gram, second), the corresponding unit of force is called a *dyne*. Since the acceleration g due to gravity on the Earth's surface is about 980 centimeters per second, the downward gravitational force exerted by a mass of one gram is about 980 dynes.

Newton's third Law is illustrated by a jet engine. The total force pushing the fuel molecules backward is equal to the force pushing the jet plane forward.

The second and third laws together imply that in a system free of external forces, total momentum is conserved. For example, when a space craft ignites its propulsion jet, its increase in forward momentum is equal to the total momentum of the jet fuel molecules thrown in the opposite direction. If two astronauts on a space walk start tossing a baseball back and forth, conservation of momentum will cause them to drift slowly away from each other, making their return to the space ship difficult or impossible.

Newton's laws raised the science of mechanics to a new level. He became a celebrity, due in large part to the writings of the philosopher Voltaire (1694–1778), a leader of the eighteenth-century intellectual movement known as the Enlightenment. In 1737, Voltaire published a commentary on Newton's *Principia*.

There were important matters that eluded even Newton. For example, Newton did not define kinetic energy, a concept introduced by Gottfried Leibniz, which he called *vis viva*, "living force."[3] In 1740, Émilie du Châtelet (1706–1749)—a lady of the French court who became Voltaire's mistress and, more important, the distinguished scholar who translated Newton's *Principia* from Latin to French—proposed that vis viva could be converted in part to *vis mortua*—a precursor of potential energy—and that the sum of *vis viva* and *vis mortua* would remain constant. This foreshadowed the more general principle of *conservation of energy*.

Energy is a more fundamental concept than momentum because it is meaningful, not only in mechanics, but also in all branches of physics. The energy of motion of a particle is proportional to the square of its velocity. More specifically, in the *cgs* system, the kinetic energy in *ergs* of a moving particle with mass m grams and velocity v in centimeters per second is equal to $mv^2/2$. Because the time needed to brake an auto is roughly proportional to its energy, doubling the speed quadruples the time needed to brake. Châtelet was inspired by the work of a Dutch researcher, Willem 'sGravesande (1688–1742), who noticed that a ball thrown into soft clay penetrated four times as far if it had twice the speed.

10.2. ROTATIONAL DYNAMICS

Planetary motion is rotational. This section discusses how Newton's Laws can be adapted to the problems of rotational motion. Specifically, the concept of *angular momentum* will be discussed. First, we look at the simplest instance of rotational motion—uniform circular motion. We will see why it is difficult to maintain one's balance when walking on a moving merry-go-round.

Uniform circular motion

An important goal of this chapter is to study the forces governing the orbital motion of the planets. In preparation for this task, we look at the simpler problem of the forces involved in uniform circular motion.

Figure 10.2 illustrates the uniform movement of a particle from P to Q on a circular path C of radius r. Here a particle of mass m starting at P moves to Q through the angle $\Delta\theta^4$ in time Δt with a constant angular velocity ω. Hence, the angle $\Delta\theta$ is equal to $\omega\Delta t$. Using radian measure for angles, the length of the arc swept out is

$$\widehat{PQ} = r\Delta\theta = r\omega\Delta t \tag{10.1}$$

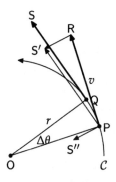

Figure 10.2 Finding the acceleration of uniform circular motion. In the time interval Δt, a particle of mass m on circular path C with center at O moves from point P to point Q through angle $\Delta\theta$ with constant speed v and constant angular velocity $\omega = v/r$.

To find the force required to produce uniform circular motion, Newton's second law says that we must find the acceleration of the particle, the rate of change of the velocity vector. Although the speed

$$v = r\omega \tag{10.2}$$

is constant, the acceleration of the velocity vector is not zero. For example, a passenger on a merry-go-round moving with constant angular velocity experiences an outward pull due to the force resulting from such a nonzero acceleration.

In Figure 10.2, the velocity vectors at P and Q are $\overrightarrow{PR}$ and $\overrightarrow{QS}$, respectively, both with length v. To subtract these vectors, they should have a common origin. To this end, vector $\overrightarrow{PS'}$ is equal to vector $\overrightarrow{QS}$, and the difference $\overrightarrow{PS'} - \overrightarrow{PR}$ is equal to $\overrightarrow{RS'}$. It is convenient to move this vector to $\overrightarrow{PS''}$ so that its tail is at point P. The figure shows that $\overrightarrow{PS''}$ very nearly points to the center of the circle O. In fact, it can be shown that as $\Delta\theta$ tends to zero, the direction of $\overrightarrow{PS''}$ tends to the direction of the radius $\overrightarrow{PO}$.

Using radian measure for angles, $r\Delta\theta$ is equal to the length of the circular arc PQ. The triangles $\triangle PRS'$ and $\triangle OPQ$ are isosceles and similar. Therefore,

$$\frac{\overline{PQ}}{r} = \frac{\overline{RS'}}{v} \tag{10.3}$$

But the arc length of PQ is close to length of the line segment PQ, and thus

$$\Delta\theta \approx \frac{\overline{RS'}}{v} \tag{10.4}$$

The average acceleration vector as the particle moves from P to Q in time Δt is equal to the change in velocity in time Δt

$$\frac{\overrightarrow{QS} - \overrightarrow{PR}}{\Delta t} = \frac{\overrightarrow{RS'}}{\Delta t} \tag{10.5}$$

Using (10.4), the magnitude of this average acceleration vector is approximately

$$v\frac{\Delta\theta}{\Delta t}$$

It can be shown that the above approximation is sufficient to ensure that, as Δt tends to zero, the magnitude of the acceleration tends to

$$v\frac{d\theta}{dt} = v\omega$$

Using (10.2), the acceleration is also equal to

$$r\omega^2 = \frac{v^2}{r} \tag{10.6}$$

The direction of the acceleration vector is the limiting direction of $\overrightarrow{PS''}$, which, as noted above, coincides with the direction of the radius vector $\overrightarrow{PO}$.

Thus, according to Newton's second law, in order to maintain uniform circular motion, a centripetal force is required of magnitude mv^2/r.

Angular momentum

The dynamics of rotation has application not only to the motions of the solar system but also to more mundane phenomena, from tornados to the pirouettes of ballet dancers; from a spinning top to the vortex of a flushing toilet. Is it true that the direction of spin of water in a flushing toilet depends on the hemisphere in which the toilet is located? Why is it that an ice skater can increase his or her rate of spin by moving arms and legs closer to the axis of spin?

In this section, the concept of angular momentum is defined. Newton discovered that conservation of angular momentum is equivalent to Kepler's second law.

Consider an arbitrary moving object in relation to a fixed axis. For example, the object could be a planet and the axis could be the Sun. The simplest case, shown in Figure 10.3, is a single particle P with constant mass m moving in a two-dimensional plane, and the axis is the fixed point O. Let r be the radial distance $\overline{OP}$, and let θ be the angle between OP and a fixed half-line l emanating from O. The angle is considered positive or negative, depending on whether its sense is counterclockwise or clockwise. Moreover, ω denotes the angular velocity of P relative to O—that is, $\omega = d\theta/dt$, the rate of change of the angle θ.

The *angular momentum L* of the particle P, analogous to linear momentum mentioned in Newton's second law of motion, is defined by the formula

$$L = mr^2\omega \qquad\qquad (10.7)$$

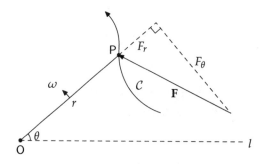

Figure 10.3 Rotational dynamics of particle P with respect to axis O. The particle P with a constant mass m moves on a curvilinear path C and is subjected to force **F**. The angular momentum L of particle P is defined as $mr^2\omega$. The *torque* T is defined as the angular component F_θ of force **F** times the radius: $T = rF_\theta$.

the mass times the square of the radial distance times the angular velocity. The name angular *momentum* is used because, as will be seen, angular momentum for rotational motion is analogous to linear momentum mentioned in Newton's second law of motion in Section 10.1.

In this section, propositions concerning angular momentum assume the motion to be of a *single particle* in a *two-dimensional plane*. Although angular momentum can be considered in greater generality, this is the simplest model of a planetary orbit.

Suppose that a force **F** acts on particle P, as shown in Figure 10.3. Force, a vector quantity, has both magnitude and direction. The force **F** is completely described by giving its *components*. Figure 10.3 shows the force **F** together with its radial and angular components, F_r and F_θ, respectively. The *torque* T with respect to O produced by the force **F** applied at a point P is defined as the distance $\overline{OP}$ between the axis O and the point P multiplied by the angular component of the force—that is, $T = rF_\theta$. Note that a centripetal force, a force of attraction or repulsion directed radially along OP, produces zero torque on the particle at P. Gravitation is an example of a centripetal force. A suitably applied centripetal force can produce a wide variety of paths, including the one shown in Figure 10.3. Any possible centripetal force is ignored in Figure 10.3 because such a force has no effect on torque.

It can be shown using Newton's Laws of motion that torque T is equal to the rate of change over time of the angular momentum L. The relationship is analogous to Newton's second law, which states that force is equal to the rate of change of *linear* momentum.

Figure 10.3 can be modified to include more than one particle. The total torque is the sum of torques corresponding to each particle. Forces directed radially or along a line connecting any two particles can be ignored in computing total torque. The total torque is equal to the rate of change over time of the total angular momentum.

Proposition 10.1 (Conservation of angular momentum). *If total torque is zero, then the total angular momentum of the system is constant.*

Conservation of angular momentum is exhibited when a spinning ice skater increases the rate of spin by moving his or her arms and legs closer to the axis of spin.

Figures 10.4a and b show the path of the single particle in Figure 10.3 without any torque acting on it. (There can be a radial force of attraction or repulsion, but as previously noted, such a force produces no torque and has no effect on angular momentum.) Thus, the moving particle in these figures must conserve angular momentum. In other words, using equation (10.7), although r and ω may both vary with time, the product $L = mr^2\omega$ must remain constant.

Proposition 10.2. *The angular momentum $L = mr^2\omega$ of a particle of mass m moving in a two-dimensional plane with respect to an axis at O is also given by the formula*

$$L = 2m\frac{dA}{dt} \tag{10.8}$$

where dA/dt is the rate at which the radius from O to the moving particle traces out area.

For example, in Figure 10.4a, if the particle moves from P to Q and from Q to R in equal time intervals Δt, then the two areas labeled $\mathcal{A}$ are equal.

Proof of Proposition 10.2. The following argument is intended to establish plausibility. A rigorous mathematical argument would require some more careful assumptions concerning the smoothness of the curve $\mathcal{C}$ and detailed handling of error estimates.

In Figure 10.4b, two nearby points P_1 and P_2 on the path $\mathcal{C}$ are connected to O. Apart from the fact that P_1P_2 is not a straight line segment, OP_1P_2 is a slender triangle. The dashed curve $\mathcal{D}$ is an arc of the circle with center at O and radius $r_1 = \overline{OP_1}$. The shaded region OP_1Q is a sector of circle $\mathcal{D}$ with centripetal angle $\Delta\theta$ measured in radians (1 radian $= 180/\pi$ degrees). The ratio of the area of the sector OP_1Q to the area of the entire circle $\mathcal{D}$ is equal to the ratio of the angle $\Delta\theta$ to the angular measure of the entire circle 2π. It follows that area of the sector is equal to the area of the entire circle πr_1^2 times $\Delta\theta/2\pi$. Putting ΔA equal to the area of the shaded region,

$$\Delta A = \frac{1}{2}r_1^2\Delta\theta$$

(Angles in degrees would make this formula slightly more complicated.)

The area ΔA of the shaded sector OP_1Q is an approximation for the area of region OP_1P_2 with error equal to the area of the small region QP_1P_2. If the angle $\Delta\theta$ is halved, then the area of the shaded sector is halved and the area of error triangle QP_1P_2 is divided by four. In other words, using the shaded sector OP_1Q as an approximation

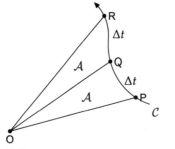

(a) Equal areas $\mathcal{A}$ are swept out in equal time intervals Δt.

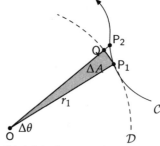

(b) ΔA is the area of the shaded sector, subtending the angle $\Delta\theta$.

Figure 10.4

for the area OP_1P_2, if the angle $\Delta\theta$ is halved, then the *per cent* error is approximately halved.

Thus if the particle travels from P_1 to P_2 in time Δt while the radius from O sweeps out the area ΔA, the average rate of change of this area is given approximately by

$$\frac{\Delta A}{\Delta t} = \frac{1}{2}r_1^2\frac{\Delta\theta}{\Delta t}$$

From the argument in the paragraph above, the percentage error of this approximation tends to zero at a geometric rate. Mathematicians of Newton's time would say that if angle $d\theta$ and area dA correspond to the *infinitesimal* interval of time dt, then the *instantaneous rate* dA/dt at which area swept out when the particle is at point P_1 is *exactly*

$$\frac{dA}{dt} = \frac{1}{2}r_1^2\frac{d\theta}{dt} \tag{10.9}$$

□

If a particle moves without any applied torque, then Propositions 10.1 and 10.2 imply that area is swept out at a constant rate. But the condition prescribing equal areas in equal times is precisely Kepler's Second Law! Thus, Kepler's Second Law is equivalent to the assertion that a planet moves so that its *angular momentum with respect to the Sun is constant*.

Propositions 10.1 and 10.2 show that *any* centripetal force of attraction or repulsion would produce a motion that satisfies Kepler's second law. Kepler's second law does not determine the inverse-square law of gravitational attraction because momentum is conserved by any centripetal force. However, we will see that the Kepler's first law, which asserts that planetary orbits are ellipses with the Sun at a focus, gives the additional strength needed to imply the inverse-square law.

A version of Proposition 10.2 that will be needed later is illustrated in Figure 10.5 in which a particle moves on path C. We will determine a new formula for the angular momentum with respect to point O. At point P, the radial distance is $r = \overline{OP}$, the tangent is line T, which determines the direction of the velocity vector $\overrightarrow{PR}$. The speed, the magnitude of the velocity vector PR, is denoted p. The component of the velocity in the direction of rotation about O is $q = PS$. The line through O perpendicular to the tangent meets T at point Q. The distance from tangent T to the axis O is denoted $s = \overline{OQ}$.

The angular velocity ω at P is given by $\omega = q/r$. The triangles $\triangle OPQ$ and $\triangle PRS$ are similar. Hence, $q/p = s/r$. Thus, using (10.7), angular momentum is given by

$$L = mr^2\omega = mr^2(q/r) = mrq = mps$$

The following proposition summarizes the above result:

Proposition 10.3. *Suppose that a particle (mass $= m$) at point P moves with respect to axis O. As in Figure 10.5, assume that p, q, r, and s are, respectively, the speed, the angular component*

of velocity, the radial distance, and the distance to the tangent line. Then the angular momentum L
is given by the equation

$$L = mrq = mps \tag{10.10}$$

The Coriolis effect

Newton's first law of motion, the Law of Inertia, asserts that a particle at rest remains
at rest, and a particle in motion continues to move in a straight line with a constant
speed unless an external force acts upon it. The Law of Inertia is correct for an observer,
called an *inertial observer*, whose frame of reference does not accelerate—a condition
that excludes rotation. Any deviation from the Law of Inertia observed from a rotating
frame of reference is called a Coriolis effect, named after the French mathematician
Gaspard-Gustave Coriolis (1792–1843), who wrote on this topic. Newton formulated
the Law of Inertia despite living on a rotating planet because, over sufficiently short
distances and slow-enough speeds, the Coriolis effect is negligible.

The Coriolis effect is too small to affect the flow in toilet bowls, but it has a major
influence on atmospheric and oceanic currents because they move great distances.
The Coriolis effect produces cyclonic atmospheric effects, clockwise in the northern
hemisphere and counterclockwise in the southern. Thanks to the Coriolis effect, the
world's weather is not static. The Coriolis effect also explains the magnetic poles, which
are the result of cyclonic movement of charged particles in the molten metallic mass
deep in the Earth's interior.

In the northern hemisphere, the Coriolis effect deflects a ball thrown in any direction
slightly to the right.[5] A ball thrown to the east veers south because it tends to follow
a great circle rather than a circle of latitude. A ball thrown north veers east because of
conservation of angular momentum. Rather than elaborate these two ideas, it is simpler
to derive the Coriolis effect in the following two steps. (1) Determine the motion of an
object from the point of view of an inertial observer. (2) Translate the motion as seen

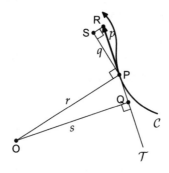

Figure 10.5

by an inertial observer to the frame of reference of a rotating observer. The following two sections use this method to analyze the Coriolis effect for two examples—the lost marble and the slipping astronaut.

The lost marble

Problem 10.1. Sue, a child on a merry-go-round, accidentally dropped a marble. Find the path of the marble.

Let us make a few assumptions:

1. The merry-go-round rotates clockwise with constant angular velocity.
2. The initial position of the marble is midway between the center and the circumference of the merry-go-round.
3. The marble moves with negligible friction, and the dynamics of rotation of the marble can be ignored. We approximate the marble as a particle that slides without friction.
4. The initial velocity of the marble is zero relative to the moving merry-go-round.
5. The marble rolls freely without hitting any obstacles.

Figure 10.6 superimposes two views of the moving marble—(1) the view seen by Sue's father, Art, an inertial observer[6] who does not ride the merry-go-round, and (2) the view seen by Sue, who is a rotating observer. Art sees the merry-go-round moving clockwise. Sue sees the merry-go-round as stationary and rest of the world, including Art, as moving counterclockwise.

The circle $\mathcal{M}$ is the outer edge of the merry-go-round, and $\mathcal{N}$ is a circle concentric to $\mathcal{M}$ with half its radius. Sue drops her marble at P, a point on $\mathcal{N}$. The line $\mathcal{L}$ is tangent to $\mathcal{N}$ at P. Art is located at A, a point outside of $\mathcal{M}$ on the extension of the tangent line $\mathcal{L}$.

Initially, the marble has the velocity of the turning merry-go-round directed along the line $\mathcal{L}$. Since P is midway between the center O and the boundary circle $\mathcal{M}$, the speed of P is initially one-half of the speed of points on the circle $\mathcal{M}$. Art, the inertial observer, sees the marble continue directly toward him on the straight line $\mathcal{L}$ at a constant speed—in accordance with Newton's Law of Inertia.

Suppose that Art flashes a beam from his flashlight six times, at equal time intervals, along the line $\mathcal{L}$ —once, initially, with the marble at P and, thereafter as the marble reaches each of the points Q, . . . , U.

As a rotating observer, Sue sees the marble travel along the floor of the merry-go-round on the path C. To understand the construction of C, we translate the line $\mathcal{L}$, the path of the marble as seen by Art, the inertial observer at A, into the path seen by Sue and other rotating observers. Sue sees the first flash from Art's flashlight illuminate the line $\mathcal{L}$ and light up the marble at P. When Art flashes the beam the second time, Sue has rotated to a new position. From her point of view, Art has rotated counterclockwise to point A′, and she sees the second flash along line $\mathcal{L}'$ light up the marble at Q′ and graze the circle $\mathcal{N}$ tangentially at point Q″. Similarly, the four remaining flashes light up the marble at points R′, . . . , U′ and meet $\mathcal{N}$ tangentially, respectively, at R″, S″, T″, U″. The path of the marble, the goal of Problem 10.1, is the curve $\mathcal{C}$ found by linking the points P, Q′, R′, . . . , U′.

1. Remarks.

The construction of the curve $\mathcal{C}$ does not depend on the value of the constant rate of rotation of the merry-go-round. The length of the arc $\overset{\frown}{PU}''$ is equal to the length of the line segment $\overline{PU}$ because the circle $\mathcal{N}$ rotates at the same speed that was initially transferred to the marble at point P. The angle subtended at the center O by the arc $\overset{\frown}{PU}''$ is the angle that the merry-go-round has rotated during the time that the marble completes its journey along the curve $\mathcal{C}$.[7]

The next section analyzes an example of the Coriolis effect due to the rotation of Earth. Borrowing an image from McIntyre(2000),[8] imagine that Earth is perfectly spherical and coated with slippery ice providing no friction whatever. As seen below, due to the Coriolis effect, an object initially at rest proceeds on a curved path.

The slipping, sleeping astronaut

Five billion years from now, Pi, an astronaut from another galaxy, visits our planet, now reduced to a slippery icy sphere. He lands at 60° North 11° East, where once was

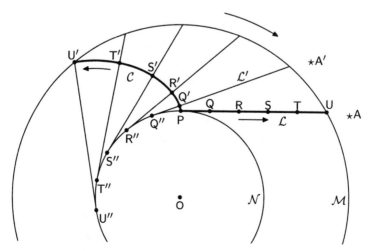

Figure 10.6 The marble on the merry-go-round.

the city of Oslo, Norway—point P in Figure 10.7. Tired from his long journey, the astronaut puts his foam pad and sleeping bag on the ice, takes off his boots, and enjoys twelve hours of uninterrupted sleep. When Pi removes his hobnail boots, which grip the ice, he has lost any control of his movement in this zero-friction environment. From that moment, zero friction implies that the rotation of the planet has no effect on his subsequent motion, which depends only on the initial velocity that he received from the rotating Earth at the moment that he removed his boots.

To discover Pi's subsequent motion, it is simplest to adopt the point of view of an observer who does not rotate with the Earth, an inertial observer named IO. In contrast, we Earthlings live our lives as rotating observers. The inertial observer uses the fixed stars for his orientation. He is guided by a celestial globe where the rotating observer uses a terrestrial globe. In order to keep both observers in mind, think of two globes—a transparent celestial globe to guide the inertial observer and a terrestrial globe rotating beneath it to guide the rotating observer. When a rotating observer, call him RO, returns home after an hour's absence, he considers that he returns to the place he left. On the other hand, from IO's point of view, Earth's rotation must be accounted for. IO observes RO returning 15° of longitude to the east of where he started. (Here and in the rest of this example, *hour* means *sidereal hour*. Earth makes a complete rotation with respect to the fixed stars in a sidereal day, 24 sidereal hours. The current value of a sidereal hour is 0.997 ordinary hours, but at the time of Pi's landing, a sidereal hour may be quite different.)

In analogy to the previous problem, the marble on the merry-go-round, the analysis of Pi's motion proceeds in two steps. (1) Chart Pi's motion as IO would see it, and (2) translate that chart to the point of view of RO. Step 2 is achieved as follows. Move each point of the inertial observer's chart westward by a suitable amount. In particular, if a point on the inertial observer's chart represents the position of an object t hours after the start of its motion, then—remembering that Earth rotates eastward at the rate of 15° per hour—that point must be moved $15t$ degrees westward in longitude for placement on the rotating observer's chart.

Newton's Law of Inertia needs some modification. On a nonrotating slippery planet, an object that receives an initial shove continues with constant velocity on a path that coincides with a taut string held on the surface of the ice.[9] Such a path is a great circle on the spherical planet. Meridians of longitude are examples of great circles, but a circle of latitude, other than the equator, is not a great circle.

Zero friction implies that the subsequent motion of a sliding object, after its initial velocity has been established, follows a great circle, *even if the planet rotates*. However, this is true only for an inertial observer—an observer who does not rotate with the planet. For such an observer, the stars are stationary, whereas an observer who rotates with the planet sees the stars move across the sky from east to west.

Figure 10.7 shows how IO sees Pi move on a different path from his boots, which are locked to the ice by their hobnails. Boots and astronaut start together at point P. However, after two hours, the amount of time in which Earth rotates 30°, the boots, following the circle of latitude $\mathcal{L}$, have moved to point Q″, but Pi, following the great circle $\mathcal{C}$, has moved to point Q.

As noted above, when Pi removes his boots he receives an initial velocity eastward— the eastward velocity of rotation of a point on the planet at 60° North. Pi's subsequent motion follows the great circle with an initial eastward direction. As shown in Figure 10.7, the circle $\mathcal{L}$ of latitude 60° N and the great circle $\mathcal{C}$ are tangent at Pi's initial position P.

The rotating observer, RO, sees Pi's movement quite differently—relative to the fixed globe rather than the rotating one seen by IO. As shown in Figure 10.7, RO sees Pi's boots fixed at point P and he sees Pi slipping along curve $\mathcal{D}$, which is related to $\mathcal{C}$ (Pi's path according to the IO) as follows. After two hours the planet has rotated 30° to the east. IO locates Pi at point Q, but RO, who does not perceive the eastward rotation, believes that IO has located Pi too far east by 30° and locates Pi at point Q′. Similarly, after four hours RO locates Pi at point R′, 60° west of R, the location chosen by IO, and so on.

When Pi awakens after his twelve-hour sleep, his boots are nowhere in sight. In fact, he has slipped along a curved path from Norway across the Atlantic all the way

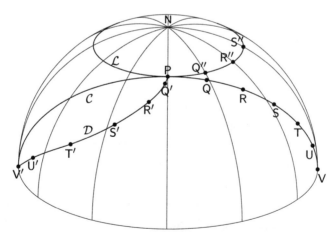

Figure 10.7 The sliding astronaut starts at point P on $\mathcal{L}$, the circle of latitude 60° North. The curve $\mathcal{C}$ is the great circle tangent to $\mathcal{L}$ at P. As seen by an inertial (nonrotating) observer, the curve $\mathcal{C}$ is the path of the sliding astronaut. Points Q, R, . . . , V mark his position at two-hour intervals. Meanwhile, the corresponding positions of his boots on the circle of latitude $\mathcal{C}$ are Q″, R″, S″, . . . As seen by an observer who rotates with the planet, the path of the drifting astronaut is $\mathcal{D}$. The astronaut's position at two-hour intervals are the points Q′, R′, . . . , V′. For the rotating observer, the astronaut's boots remain fixed at point P. Since Earth rotates eastward at the rate of 15° per sidereal hour, points Q′, R′, . . . , V′ are obtained by moving the points Q, R, . . . , V westward—Q by 30°, R by 60°, S by 90°, T by 120°, U by 150°, and V by 180°.

to Quito, Ecuador, 5,400 nautical miles from his initial location. Without his hobnail boots, Pi has little choice but to continue slipping. Will he continue slipping until he dies of exposure, thirst, and hunger? Before answering this question, consider some consequences of Pi's initial landing site at latitude 60°.

It can be shown that the radius of circle $\mathcal{L}$ is exactly half the radius of $\mathcal{C}$.[10] Consequently, the speed of a point on the Earth's surface at latitude 60° is just half of the speed of a point on the equator. Thus, Pi's initial speed, from IO's viewpoint, is exactly half of the speed of a point on the equator. Consequently, in two sidereal days, Pi will have made the entire circuit of the great circle $\mathcal{C}$, and his boots will have made the circuit of $\mathcal{L}$ exactly *twice*. At this time, Pi and his boots will be reunited. This happy reunion takes place because of Pi's fortunate choice of his initial landing site at latitude 60°.

At the moment that Pi encounters his boots, his slipping speed slows momentarily to zero. Unless he dons his boots at this time, he will repeat his slipping journey, which began two days ago.

During his twelve-hour sleep, from IO's viewpoint, Pi has drifted *east* to point V on the equator. From RO's viewpoint, Pi has drifted *west* to a point V' on the equator. Points V and V' appear to be opposite endpoints of a diameter, but, in reality, they are the same point seen from different points of view.

The Foucault pendulum

The Foucault pendulum, named after the French physicist Léon Foucault (1819–1896), is a demonstration of the Coriolis effect that can be seen in some museums and planetariums—a pendulum suspended 40 feet or more and with a bob weighing 50 pounds or more. The pendulum gives dynamical proof of the rotation of the Earth by precessing clockwise in the northern hemisphere and counterclockwise in the southern. The rate of precession varies with the geographical location and is determined by the Coriolis effect. The mathematical derivation is beyond the scope of this book.

Many years ago, as a student assistant in a physics class, I helped with a demonstration of a Foucault pendulum that hung from the ceiling of Bridges Auditorium at Pomona College—a large auditorium seating about 2,500 under a domed ceiling more than 40 feet high. The cable supporting the Foucault pendulum hung through a hole at the apex of the dome, continuing farther up about 10 feet to a point of attachment on an outer dome. As the class gathered below around the pendulum, I found my way above through the dark and dusty space between the inner and outer domes. Finally, I climbed a ladder to the highest point and switched on a mechanism to ensure that the oscillations of the pendulum did not diminish over time. This mechanism counteracted frictional effects, such as the aerodynamic resistance of the pendulum, by shortening and lengthening the supporting cable once during each oscillation of the pendulum, simulating the "pumping" of a child on a playground swing. This periodic lengthening and shortening of the cable did not distort the natural precession of the pendulum.

10.3. THE LAW OF UNIVERSAL GRAVITATION

In the previous section, it was shown that Kepler's second law is equivalent to the principle of conservation of angular momentum. Newtonian mechanics implies that any central-force law of attraction or repulsion is consistent with this principle. Newton showed that Kepler's first law, elliptical orbits with the Sun as a focus, implies a centripetal force of a particular form, the *inverse square law*.

Gravitational attraction is the earliest theory of *action at a distance*—a type of inter-action that, in another context, Einstein called "spooky."

The inverse-square law can be observed in other contexts. For example, the brightness—the *luminance*—of a point source of light, which can be measured with a photographer's light meter, diminishes according to the inverse-square law as its distance from the observer increases. Figure 10.8 shows light rays emanating from a point source at O. If the luminance at distance d is L, then the luminance at distances $2d$ and $3d$ is $L/4$ and $L/9$, respectively. This example shows that the inverse square law is not an esoteric concept. On the contrary, it stems from intuitive geometrical concepts.

In 1684, the astronomer Edmund Halley asked Newton, now a Cambridge professor, the following question: if the Sun attracted a planet with a force inversely proportional to the square of its distance, what would be the shape of the planet's orbit? Newton answered immediately, "An ellipse." The amazed Halley asked Newton how he knew this to be true. "I have calculated it," he answered. When Halley asked for details, Newton promised to provide them soon. After several months, Halley received from Newton a treatise, *De Motu*, on the subject.

This work of Newton marks the beginning of the science of *astrodynamics*, the application of dynamics to study the motion of bodies in space governed by gravitational attraction. Today, astrodynamics encompasses the engineering of orbits for artificial satellites. Astrodynamicists calculate how to control the firing of rockets to achieve desired orbits—a flight to the moon or a fly-by of Mars.

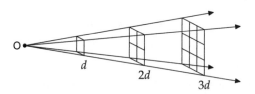

Figure 10.8 The luminance from a point source of light at O diminishes according to the inverse-square law.

Newton's reluctance to keep his colleagues informed concerning his research is typical of Newton. He discovered the calculus long before he published on this subject. One of the consequences was his unpleasant controversy with Leibniz over priority of the discovery of the calculus. Leibniz first published on this topic in 1684 whereas Newton waited until 1693. Current scholarly opinion holds that the two men discovered the calculus independently.

Newton's answer to Halley's question was not entirely accurate. It would have been more correct to say that the inverse-square law of attraction implies that the orbit is a conic section, an ellipse, parabola, or hyperbola, with a focus at the Sun. A definition is in order:

Definition 10.1. A particle moves in a *Kepler orbit* with respect to a point of attraction F if

1. the path of the motion is a conic section with a focus at F, and
2. the rotational momentum with respect to F is constant.

On July 5, 1687, Newton published his masterpiece, *Principia*, a greatly expanded version of *De Motu. Principia* contains the foundations of classical mechanics, including the law of universal gravitation. Here, Newton shows that a Kepler orbit can be attributed to the inverse-square law of attraction toward the focus F. This assertion is generally called the "direct" problem, although it is the converse of Newton's original assertion to Halley. (In *Principia*, Newton dealt not only with ellipses, but also with the other conic sections—parabolas and hyperbolas.) Newton also claimed in *Principia* to have proved the proper generalization of his original assertion to Halley—generally known as the "inverse" problem—that the inverse-square law implies the orbits are conic sections.[11] However, it has been claimed (Weinstock, 2000) that Newton did not make good this claim—more about this later.

Translations of *Principia* make the work more accessible, but its geometric style remains an obstacle. Newton chose the language of classical geometry instead of analysis, although the latter was already widely used by scholars in Newton's time. Analysis, the standard language today of mathematical physics, makes free use of algebraic notation and manipulation—both for exposition and for proofs.

The connection between Kepler's laws and the inverse-square law of gravitational attraction has been reworked by many authors. Richard Feynman's elementary treatment is presented in *Feyman's Lost Lecture* (Goodstein & Goodstein, 1996). The Irish mathematician Sir William Rowan Hamilton (1805–1865) saw a simplification in understanding this matter through a device that he called the *hodograph* (Hamilton, 1847).

The direct problem

This section gives a proof of the direct problem in the case of elliptical orbits.

The hodograph

Figure 10.9 illustrates the concept of the hodograph. Figure 10.9a represents the motion of a particle along curve C from point P to S. The velocities of the particle are represented by tangent arrows representing speed and direction at points P, Q, R, and S.

In Figure 10.9b, the points P', Q', R', and S' are located by transporting the arrows in Figure 10.9a parallel to themselves and putting their tails at a common *initial point* O. If we imagine the rest of the velocity vectors transported in this way, the heads of the arrows trace a curve $\mathcal{H}$ containing points P', Q', R', and S'. The curve $\mathcal{H}$, together with the initial point O, is called the *hodograph* of the motion shown in Figure 10.9a.

Proposition 10.4 requires a slight generalization of the hodograph concept. Instead of moving the velocity vectors parallel to themselves, as in the paragraph above, suppose those vectors are all rotated by a fixed angle a. Then we call the resulting curve, together with its initial point, an a-*hodograph*. Figure 10.9c shows a 90°-hodograph of the curve C in 10.9a.

Hamilton used a hodograph to give an alternative proof of Newton's result that Kepler's first two laws imply that a planet is attracted to the Sun by a force inversely proportional to the square of the distance between the Sun and the planet.

Proposition 10.4. *Suppose that the motion of a planet satisfies Kepler's first two laws:*

1. *The orbit is an ellipse with the Sun at one of the foci, and*
2. *the angular momentum with respect to the Sun is constant.*

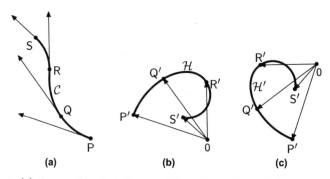

(a) (b) (c)

Figure 10.9 In (b), the curve $\mathcal{H}$ is the hodograph of the motion on the curve C shown in (a), and, in (c), $\mathcal{H}'$ is a 90°-hodograph of C.

Then it follows:

a. *Apart from a rotation of ±90° and a change of scale, the focal circle centered at the Sun is a hodograph with initial point at the second focus, the empty focus, of the ellipse.*

b. *The planet is subject to a force of attraction toward the Sun that obeys the inverse-square law.*

Proof of a. Figure 10.10a is an elaboration of Figure 7.4a. Section 7.1 contains a discussion of the use of the focal circle $\mathcal{F}$, with center at focus E and radius equal to the major axis $2a$, to determine the tangent at an arbitrary point P on the ellipse. The line connecting focus E to P intersects the focal circle at G. In Figure 10.10a, the required tangent $\mathcal{T}$ through P is the perpendicular bisector of FG.

Figure 10.10a represents a planet on an elliptical orbit about the Sun E. The velocity of the planet at point P is the vector **v** directed along the tangent line $\mathcal{T}$. Let v denote the magnitude of the vector **v**.

By Proposition 10.3, the angular momentum L with respect to the Sun E is given by

$$L = mvs \tag{10.11}$$

where m is the mass of the planet, v is its speed (the magnitude of the velocity **v**), and s is the perpendicular distance from the Sun E to the tangent line $\mathcal{T}$. Kepler's second law, together with Proposition 10.2, implies that the angular momentum L is constant.

By Proposition 10.3, the products of the distances of the foci from the tangent line $\mathcal{T}$ is equal to the square of the semiminor axis of the ellipse—that is,

$$qs = b^2 \tag{10.12}$$

From equations (10.11) and (10.12), one can see that v is proportional to $\overline{FG}$. In fact, we have

$$v = \frac{L}{ms} = \frac{Lq}{mb^2} = \frac{L}{2mb^2} \cdot \overline{FG} = k\,\overline{FG} \tag{10.13}$$

where

$$k = L/(2mb^2) \tag{10.14}$$

is constant—in part because angular momentum L is conserved.

Moreover, FG is perpendicular to the velocity vector PQ. Since the movement of the planet is counterclockwise, $\mathcal{F}$ is a hodograph, apart from a rotation of 90°

(counterclockwise because the movement of the planet is counterclockwise) and a change of scale corresponding to the constant k.

Proof of b. Figure 10.10b shows the movement of the planet along the ellipse $\mathcal{E}$. Let Δt be the time interval in which the planet travels from P to Q, subtending the angle $\Delta\theta$ at E. The corresponding points on $\mathcal{F}$, the focal circle centered at E, are R and S, respectively. Since, as shown above, $\mathcal{F}$ is a hodograph (apart from rotation and change of scale) with initial point at F, the speeds at P and Q are equal to $k\overline{FR}$ and $k\overline{FS}$.

As the planet moves from P to Q, we will see that the arc RS gives different information depending on whether it is viewed from focus F or E. Viewing RS from focus E or focus F yields information concerning *angular velocity* or *acceleration*, respectively.

Viewing RS from focus E. In Figure 10.10b, the angle $\Delta\theta$ measures the angular movement of the planet as it moves from P to Q in the time interval Δt. The average angular velocity $\overline{\omega}$ is given by

$$\overline{\omega} = k\,\frac{\overline{RS}}{2a\,\Delta t}$$

where a is the semimajor axis of the ellipse. The limit of this fraction as Δt tends to zero is the instantaneous angular velocity ω, also denoted $d\theta/dt$.

Viewing RS from focus F. The focal circle $\mathcal{F}$ is a hodograph (apart from rotation and change of scale) with initial point F. Hence, after a counterclockwise rotation by 90°, the vectors $\overrightarrow{FR}$ and $\mathbf{u} = \overrightarrow{FS}$ are proportional to the velocity vectors of the planet at positions P and Q, respectively. The subtraction of these two vectors yields

$$\overrightarrow{FS} - \overrightarrow{FR} = \overrightarrow{RS} = \Delta\mathbf{u} \tag{10.15}$$

(Note that the directed line segment $\overrightarrow{RS}$ fits closely, but not exactly, on the arc RS.) Thus $k\,\Delta\mathbf{u} = \overrightarrow{RS}$ is the change in the velocity vector in time Δt as the planet moves

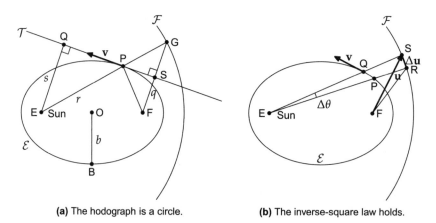

(a) The hodograph is a circle.   (b) The inverse-square law holds.

Figure 10.10 Proof of Proposition 10.3.

from P to Q, apart from a counterclockwise rotation of 90°. The rate of change in velocity is called *acceleration*. Thus the average acceleration over the time interval Δt is $k\,\Delta u/\Delta t$, and the limit as the time interval Δt tends to zero is the instantaneous acceleration vector $kd\,u/dt$. Taking account of the rotation by 90°, we see that the acceleration vector points toward the Sun E.

The magnitude of the average acceleration over the time interval Δt is $k\overline{SR}/\Delta t$. Putting the semimajor axis of the ellipse $\mathcal{E}$ equal to a, note that the radius of the focal circle $\mathcal{F}$ is $2a$. Using radian measure for $\Delta\theta$, note that the arc length of RS is equal to $2a\,\Delta\theta$. Thus the magnitude of the average acceleration is $2ka\,\Delta\theta/\Delta t$, and the instantaneous acceleration is $2kad\,\theta/dt = 2a\omega$, where ω is the angular velocity. Using Newton's second law, the planet is subject to a force whose magnitude F is the mass m of the planet times its acceleration—that is,

$$F = 2kam\omega \tag{10.16}$$

As noted above, the acceleration vector is directed toward the Sun. Therefore, the force is also directed toward the Sun.

By definition (10.7), angular momentum L is given by $L = mr^2\omega$. Solving for ω, we have $\omega = L/(mr^2)$. Putting this in equation (10.16), we have

$$F = \frac{2kaL}{r^2} = Kr^{-2} \tag{10.17}$$

where K is a constant. In particular,

$$K = aL^2/mb^2 \tag{10.18}$$

Because a and L are constants, formula (10.17) shows that the inverse-square law is satisfied. □

Newton proved, as in Proposition 10.3 above, that Kepler's first two laws imply that the Sun exerts an inverse-square force of attraction on the planets—that is, a force proportional to $1/r^2$, where r is the distance of the planet from the Sun. Newton's law of *universal* gravitation asserts that every pair of particles in the universe is subject to a force of attraction directed along the line connecting the two particles. If the masses of the particles are m_1 and m_2 and the distance between them is r, then the magnitude of the gravitational force is

$$G\frac{m_1m_2}{r^2} \tag{10.19}$$

where G is the gravitational constant.

The Inverse problem

The previous section gave a solution to the direct problem—showing that if a particle moves on an elliptical orbit satisfying Kepler's first two laws with respect to a point of attraction, then there exists a central force on the particle proportional to the inverse-square of the distance between the particle and the point of attraction. Newton also showed that the same is true for an arbitrary Kepler orbit—that is, for motion on an arbitrary conic section with constant angular momentum with respect to the point of attraction. In this section we consider the inverse problem—the converse of Proposition 10.4:

Proposition 10.5. *Suppose a particle* P *moves under the influence of an inverse-square force of attraction or repulsion with center at point* F. *Then* P *moves in a Kepler orbit (see Definition 10.1) with focus at* F.

In *Principia*, Newton devotes Propositions 11–13 of Book I to the inverse problem, but his proof is incomplete. It seems that Newton wanted to make use of the solution of the direct problem, Proposition 10.3, in order to prove the inverse problem, Proposition 10.5. This seems a bit like lifting oneself by one's bootstraps because, in general, *A implies B* does not entail *B implies A*. However, this idea does lead to success in this special case if the following two propositions can be proved.

Proposition 10.6. *Let* F *be a given point in space. Suppose that the initial position and velocity of a moving particle* P *are specified. Moreover, suppose that initial acceleration of the particle is directed toward* F *with magnitude a. Then for every positive or negative choice of a there exists a distinct Kepler orbit, with* F *as its focus, such that the particle* P *has the specified initial position, velocity, and acceleration.*

In *Principia*, Newton shows essentially this in Propositions 11–13. However, Newton specifies initial curvature of the Kepler orbit rather than its initial central acceleration.

Proposition 10.7. *Suppose an inverse-square law is given—a force of attraction or repulsion with respect to a point* Q. *Let* P *be a particle with specified initial position and velocity. Then there exists one and only one motion of the particle subject to the given central force and satisfying these initial conditions.*

The proofs of these propositions will not be given here. In particular, the proof of Proposition 10.7 is beyond the scope of this book.

In *Principia* Propositions 11–13 and in 39–41 where Newton discusses the problem of determining the path of a particle subject to a more general central force law, Newton does not deal with the uniqueness question raised above in Proposition 10.7. This is not surprising because the question of uniqueness of solutions of differential equations was not settled until about 100 years after Newton by the French mathematician Augustin Louis Cauchy (1789–1857).

Propositions 10.6 and 10.7, together with Proposition 10.3, solve the inverse problem. The reasoning is as follows:

1. Suppose a particular inverse-square law is given. We wish to show that the motion executed by a particle with given initial position and velocity is a Kepler orbit. The initial acceleration is specified by the strength of the given inverse-square law.
2. Proposition 10.6 asserts that there is a unique Kepler orbit determined by the initial position, velocity, and acceleration.
3. Proposition 10.4 asserts that the Kepler orbit above is generated by an inverse-square law, which must coincide with the given inverse-square law.
4. The proof is finished if it can be shown that the motion is unique—that no more than one motion is possible for a particle with the given initial position and velocity and governed by the given inverse-square law. But this is assured by Proposition 10.7.

Proposition 10.7 is more subtle than it might seem because there are examples of central-force laws for which the initial position and velocity of a particle do not determine a unique path. In the following example, the motion is one-dimensional.

The cube-root force of repulsion

In *Principia*, Book I, Proposition 39 gives a method of finding the motion of a falling body subject to an arbitrary central force, but Newton does not deal with the question of uniqueness of the motion, a critical issue in the solution of the inverse problem. Common sense seems to tell us that if we know the initial conditions of a particle, and if we know the nature of the force acting on this particle, then the subsequent motion of the particle is determined uniquely. Indeed, it can be shown that, for the inverse-square law, the initial position and velocity of a particle determine its subsequent motion, but this is not true for *all* central force laws. The following is an example in which a body, initially at rest, is subject to a central force that does *not* uniquely determine the body's subsequent motion.

When Galileo dropped a ball from the Leaning Tower of Pisa, the subsequent motion of the ball was uniquely determined by the nature of the force of gravity. Would this be true for any conceivable force instead of the usual gravity? The surprising answer is "No," as shown by the following example.

Consider the following thought experiment. Suppose a particle is subject to a force of repulsion proportional to the cube root of the distance from the origin. In particular, suppose the mass m is equal to 1, x is the distance of the particle from the origin, and force of repulsion is equal to $6\sqrt[3]{x}$. Suppose that initially (at time $t = 0$) the particle is located at the origin with velocity zero. It can be shown using elementary calculus that this initial value problem has more than one solution. This result is illustrated in Figure 10.11. There are infinitely many motions possible, of which five, labeled $\mathcal{A}, \ldots, \mathcal{E}$, are shown in Figure 10.11.

In Figure 10.11, the straight line $\mathcal{E}$ is the graph of a particle permanently at rest at origin—a solution of the cube-root force law of repulsion because if the particle remains at the origin the force acting on it is zero. In $\mathcal{A}$ and $\mathcal{B}$, the particle moves to the right or left, respectively, satisfying $x = \pm t^3$ where x is the distance of the particle from its position at time $t = 0$. In $\mathcal{C}$ and $\mathcal{D}$, the particle remains motionless for time intervals $0 \leq t \leq c$ and $0 \leq t \leq d$, respectively, after which it moves right or left such that $x = (t - c)^3$ or $x = -(t - d)^3$, respectively.

Thus, an initially motionless ball dropped from the origin subject to cube-root repulsion (1) can remain suspended motionless forever, (2) can move up or down immediately, or (3) up or down after an initial rest period.

A particle subject to such a force while resting at the origin is highly unstable because the slightest deflection will send it off on a path up or down. The motion illustrated here is unstable, but *nonuniqueness*, not instability, is the point of this example. In this example, the force law together with the initial conditions do not determine a unique subsequent motion. What would happen if this situation could be achieved physically? That question is not meaningful because it would be impossible to achieve the initial conditions with absolute precision. Nonuniqueness occurs only if the initial position is *exactly* at the orgin, but physical measurements always have a percentage error.

This example of cube-root repulsion warns us that there must be something different about the inverse-square initial value problem. This question is a special case of a more general question about differential equations. A central-force law is an example of a differential equation—an equation involving derivatives. What conditions ensure

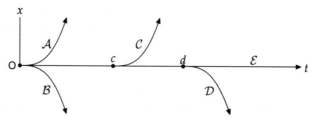

Figure 10.11 Motion along a straight line of a particle governed by a central force of repulsion proportional to the cube root of the distance from its starting point is not uniquely determined by its initial position and velocity. For a particle, initially at rest, governed by such a central-force law, the above graph shows several possible linear motions ($\mathcal{A}, \ldots \mathcal{E}$), all of which have the same initial conditions—zero velocity at the starting point O.

that a differential equation has a unique solution satisfying given initial conditions? This is not a question considered by Newton or his contemporaries. This question is relevant because Newton tells us that the inverse-square force law together with initial conditions determines the motion of a body uniquely.

Gravitational attraction of a spherical shell

In the above sections concerning the direct and inverse problems, mass points are used to model the planets and the Sun. Such an approximation seems acceptable because the Sun and the planets are much smaller than the distances between them. However, Newton showed that this proviso is unnecessary—that, regardless of their size, there is no accuracy lost in replacing the Sun and planets by particles because the gravitational attraction outside of a spherical mass is unchanged if the entire mass is concentrated at the center of the sphere. To be accurate, we should add an assumption concerning the density of the spherical mass. We will see that it is sufficient to assume the density at a point inside the sphere depends only on the distance from that point to the center of the sphere—in other words, that the Sun or planet consists of spherical shells of constant density.

Newton considers the gravitational attraction of a spherical shell with constant density. Although he did not suggest that there exist astronomical bodies with this shape, a planet or the Sun can be conceived as onionlike layers of spherical shells. Newton proved two surprising properties of gravitational shells with constant density.

Proposition 10.8. *At any point inside the shell, the gravitational attraction is zero.*

Proposition 10.9. *At any point outside a given spherical shell, the gravitation attraction of a spherical shell is the same as if the entire mass were concentrated at the center of the sphere.*

Propositions 10.8 and 10.9 played a critical role in the discovery that electric charges produce a force that also follows the inverse-square law. Benjamin Franklin (1706–1790) performed an experiment in which he observed that a sliver of cork was attracted to an electrically charged metal can, provided the cork was *outside* the can, but, curiously, no attraction was observed if the cork was *inside* the can. Franklin communicated this result to the English chemist and clergyman Joseph Priestley (1733–1804). In 1766, Priestley, who was acquainted with Newton's result concerning the gravitational attraction of spherical shells, reasoned that Franklin's experiment showed that the electric force behaved in a similar manner. Priestley cleverly surmised that the electric force, like the gravitational force, might satisfy the inverse square law. This was subsequently demonstrated in 1785 by the French military engineer and physicist Charles

Augustin de Coulomb (1736–1806). The absence of electric force inside a conducting cage, called a Faraday cage, is important in a variety of technological applications.

Propositions 10.8 and 10.9 are contained in Newton's *Principia* (Book 1, Propositions 70 and 71). His proof of Proposition 10.9 will not be discussed here. I believe that Newton's brief proof of Proposition 10.8 has a gap. The following is an attempt to repair this shortcoming. Newton's arguments use infinitesimals implicitly. Following Newton, the proof below also uses infinitesimals.

Proof of Proposition 10.8. The proof is illustrated in Figure 10.12. S is a spherical shell with uniform density, and P is an arbitrary point inside S. The dashed "circles of longitude" are intended to help visualize the sphere S.

The idea of the proof of Proposition 10.8 is to show that the surface S consists of pairs of balancing infinitesimal subregions such that the net gravitational force at P from each pair is zero. In Figure 10.12, A and A' are a pair of balancing subregions. Later in the proof, these subregions must shrink—becoming infinitesimally small.

The pairs are constructed as follows. Given a small region A on S, let A' be the projection of A through the point P. More precisely, if Q is a point in A, then its projection in A' is the point R' where the extension of the straight line QP meets the surface of the sphere S. Similarly, point R in A projects to point Q' in A'. The points

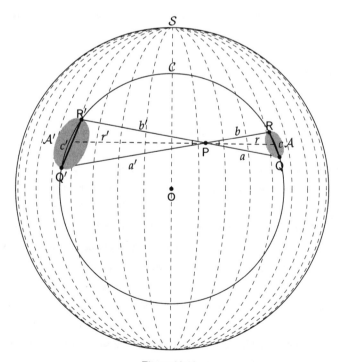

Figure 10.12

P, Q, and R determine the circle $\mathcal{C}$—the intersection of the sphere $\mathcal{S}$ with the plane containing points P, Q, and R.[12]

The next step is to find the relation between distances in $\mathcal{A}$ and their image distances in $\mathcal{A}'$. How does the projection process distort distances? The key is the fact that the triangles $\triangle$PQR and $\triangle$PQ'R' are similar. This is a consequence of the theorem concerning intersecting circular chords. The intersection point divides each chord into two segments. The theorem asserts that the product of the lengths of these two segments is the same for each chord. For example, in Figure 10.12, the chords RQ' and R'Q intersect in point P, and, consequently, $ab' = ba'$. Rewriting this equation as $a' : a = b' : b$ and noting that the angles $\angle$RPQ and $\angle$R'PQ' are equal because they are vertical angles of a pair of intersecting straight lines, it follows that the triangles $\triangle$PQR and $\triangle$PQ'R' are similar with the scale factor $\lambda = a'/a = b'/b$. Note that similarity of the triangles implies that c'/c is also equal to λ.

The linear distance $\overline{QR} = c$ is slightly smaller than the length of the circular arc QR, to be denoted $\widehat{QR}$. Similarly, $\overline{Q'R'} = c'$ is slightly smaller than the arc length $\widehat{Q'R'}$. However, if the angle $\angle$QPR is sufficiently small, the percentage errors of these approximations ($c \approx \widehat{QR}$ and $c' \approx \widehat{Q'R'}$) become as small as desired.

Newton would say that if the region $\mathcal{A}$ is infinitesimally small, then the distances a and b are equal, as are also the distances a' and b'. Let r and r' denote their common values—that is, $r = a = b$ and $r' = a' = b' = \lambda r$. The respective distances of the regions $\mathcal{A}$ and $\mathcal{A}'$ from point P are r and $r' = \lambda r$.

The images in $\mathcal{A}'$ of *linear* distances in $\mathcal{A}$ are scaled by the factor λ. It follows that areas are scaled by the factor λ^2. Thus, denoting the areas of $\mathcal{A}$ and $\mathcal{A}'$ as A and A', respectively, we have $A' = \lambda^2 A$.

The gravitational forces at point P due to the masses of $\mathcal{A}$ and $\mathcal{A}'$ are in opposite directions. According to the inverse-square law, the force on a unit mass at P due to $\mathcal{A}$ is equal to

$$\frac{GdA}{r^2} \tag{10.20}$$

where G is the universal gravitational constant and d is the surface density of the sphere $\mathcal{S}$ in units of mass per unit of area. On the other hand, the force due to $\mathcal{A}'$ is

$$Gd\frac{A'}{r'^2} = Gd\frac{\lambda^2 A}{(\lambda r)^2} = Gd\frac{A}{r^2} \tag{10.21}$$

Comparing equations (10.21) and (10.20), we see that the forces due to regions $\mathcal{A}'$ and $\mathcal{A}$ are equal in magnitude. Since they are oppositely directed, the net force at point P is zero. $\square$

Kepler's third law

Proposition 10.4 shows that Newton's inverse-square law of gravitation is implied by Kepler's first two laws. This section deals with Kepler's third law. Surprisingly, the third law is a consequence of the first two. It will be shown below that Kepler's third law is true for circular planetary orbits. Of course, actual planetary orbits are only approximately circular. However, numerous artificial satellites circle the Earth in circular orbits.

Suppose these satellites move with uniform circular motion subject to the gravitation of the Earth, which is inversely proportional to the square of the distance r from the satellite to the center of the Earth. The resulting centripetal acceleration is k/r^2 where k is a constant. But equation (10.6) gives a different representation of the acceleration based on uniform circular motion. Equating these two representations of the centripetal acceleration, we have

$$\frac{k}{r^2} = r\omega^2 \tag{10.22}$$

The period T, the length of time to complete a full circle, has an inverse relationship with angular velocity: $T\omega = 2\pi$. This leads to the following formula for the magnitude of centripetal acceleration:

$$r\omega^2 = \frac{4\pi^2 r}{T^2} \tag{10.23}$$

Thus, formula (10.22) implies $kT^2 = 4\pi^2 r^3$, or, putting $K = 4\pi^2/k$, we have

$$T^2 = Kr^3 \tag{10.24}$$

Note that K is a constant depending only on the attracting body, the Earth. This formula verifies Kepler's third law for the system of artificial satellites in circular orbits about the Earth: for any two such satellites, the ratio of the squares of their periods is equal to the ratio of the cubes of their radii.

However, the above restriction to circular orbits is unnecessary. Item III' on page 112 contains the more general statement of Kepler's third law. For any elliptical orbit, formula (10.24) remains correct if, as before, T is the period but r is now taken to be the semimajor axis of the elliptical orbit.

This chapter attempted to show how Newton brought the understanding of the motions of the solar system to a new level. He laid the foundations of the science of mechanics. He conceived of a new kind of force that moves objects without making contact with them—the force of gravity. He showed how gravity could explain the motion of the planets as well as the descent of an apple.

10.4. EPILOGUE

Today we are accustomed to the sophisticated notion that all motion is relative to an arbitrarily chosen coordinate system. However, before the twentieth century, common sense seemed to require that there should be an absolute standard for celestial motion. Ptolemy believed that Earth was motionless. On the other hand, Galileo believed so strongly that Earth orbited about the Sun that he risked and endured punishment for his belief. Galileo believed that his preference was supported by incontrovertible proof. Although this degree of confidence was unsupported by his evidence, Galileo's belief led to the crucial advances in astronomy by Kepler and Newton.

The extent of the solar system was expanded by the discoveries of the planets Uranus, Neptune, and Pluto. Uranus was discovered by William Herschel in 1781 by telescopic observation. The discovery of Neptune in 1846 proceeded from mathematical analysis of perturbations of Neptune's orbit, done independently by John Couch Adams and Urbain Le Verrier. It was hoped that observations of similar perturbations would lead to the discovery of a ninth planet. The planet Pluto was discovered in 1930 by Clyde Tombaugh, due more to persistent searching rather than mathematical calculation.

In recent years, the space program pushed planetary astronomy into the realm of *technology*. Space travel has brought the validity of planetary astronomy to a new and higher level. Newtonian mechanics and Galileo's motionless Sun are completely adequate for the foreseeable needs of space travel, whether manned or robotic. Nevertheless, 120 years ago, Newtonian mechanics did not meet a crucial test—the Michelson–Morley experiment.

In 1887, Albert Michelson (1852–1931) and Edward Morley (1838–1923) performed an experiment that changed the course of physics. Just as sound waves travel through a medium, the atmosphere, Newtonian mechanics implied that the universe must be pervaded by an invisible medium, the ether, through which light travels. Michelson and Morley devised an experiment to determine the "ether drift"—the motion of Earth through the ether.

An analogy, ascribed to Michelson himself, describes the concept of the Michelson-Morley experiment. Suppose that two swimmers, named Photon One and Photon Two, swim with exactly the same speed c. The swimming contest is located on the banks of the Ether River of width w flowing with constant speed v. Photon One swims across the river and returns to his starting point. Photon Two swims upstream a distance exactly w and returns. Who wins? An algebraic calculation involving c, w, and v shows that the cross-stream swimmer, Photon One, always wins, provided that the speed of the river is non-zero.[13]

Michelson and Morley used an interferometer, a very sensitive and accurate device for comparing the speed of light in two orthogonal directions. But, again and again,

Michelson and Morley found that the two light beams always tie. The only possible conclusion was that the speed of the river, the ether drift, must be zero—contrary to the expectation of Newtonian mechanics. Although efforts were made to "save" Newtonian mechanics (the Fitzgerald–Lorentz contraction), eventually Albert Einstein rejected the concept of ether and replaced Newtonian with relativistic mechanics.[14] Einstein replaced the concept of three-dimensional space with a four-dimensional space-time continuum. On the basis of this theory, Einstein predicted the deflection of starlight by the Sun, which was confirmed by observations of a solar eclipse in 1919 by a team led by Arthur Eddington (1882–1944).

Nineteen-nineteen was also the year in which the astronomer Edwin Hubble (1889–1953) accepted a position at the Mt. Wilson Observatory near Pasadena, California. As a result of observations in 1923–1924 with the new 100-inch Hooker Telescope, Hubble announced, on New Year's Day 1925, that he had observed objects much more distant than anything in the Milky Way. This finding expanded the size of the known universe by many orders of magnitude. In the following years, Hubble amplified his discovery through study of the spectrographic redshift. Hubble is responsible for our present-day concept of the extent of the universe.

Although planetary astronomy is by no means dead, Newton's contribution ends the story to be told here—closing the final act of the Ballet of the Planets. Hubble, on the other hand, raised the curtain for a new and ongoing theater—the Ballet of the Galaxies.

A. THE GREEK ALPHABET

A	α	Alpha
B	β	Beta
Γ	γ	Gamma
Δ	δ	Delta
E	ϵ	Epsilon
Z	ζ	Zeta
H	η	Eta
Θ	θ	Theta
I	ι	Iota
K	κ	Kappa
Λ	λ	Lambda
M	λ	Mu
N	ν	Nu
Ξ	ξ	Xi
O	o	Omicron
Π	π	Pi
P	ρ	Rho
Σ	σ	Sigma
T	τ	Tau
Y	υ	Upsilon
Φ	ϕ	Phi
X	χ	Chi
Ψ	ψ	Psi
Ω	ω	Omega

B. VECTORS

1. Equality

Vectors are considered equal if they have the same magnitude and direction. For example, in Figure B.1a the vectors $\overrightarrow{AB}$ and $\mathbf{c}$ have the same magnitude and direction, and, therefore, although the vectors are not identical, they are considered equal, and we can write $\overrightarrow{AB} = \mathbf{c}$.

2. Vector addition

Figure B.1b illustrates the addition of vectors. To add vectors $\mathbf{a}$ and $\mathbf{b}$, place the tail of vector $\mathbf{b}$ at the head of vector $\mathbf{a}$. Then $\mathbf{a} + \mathbf{b}$ is the vector whose tail coincides with that of $\mathbf{a}$ and whose head coincides with that of $\mathbf{b}$. Figure B.1b shows that the same sum is obtained if in the above procedure the order of $\mathbf{a}$ and $\mathbf{b}$ is interchanged. In either case the resulting sum is diagonal of the resulting parallelogram. Thus, we have $\mathbf{a} + \mathbf{b} = \mathbf{b} + \mathbf{a}$.

If the Moon has velocity $\mathbf{a}$ with respect to Earth and Earth has velocity $\mathbf{b}$ with respect to the Sun, then the Moon has velocity $\mathbf{a} + \mathbf{b}$ with respect to the Sun.

3. Vector subtraction

Figure B.1a illustrates the subtraction of vectors. To subtract $\mathbf{a}$ from $\mathbf{b}$, place both vectors with their tails at a common point. Then the vector from the head of $\mathbf{a}$ to the head of $\mathbf{b}$ is $\mathbf{b} - \mathbf{a}$.

4. Scalar multiplication

If $\mathbf{a}$ is a vector and k is a nonnegative scalar, then $k\mathbf{a}$ is defined as a vector with the same direction as $\mathbf{a}$, and with magnitude equal to the magnitude of $\mathbf{a}$ multiplied by k.

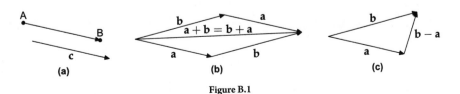

Figure B.1

The vector $-\mathbf{a}$ is defined to be the vector with the same magnitude as $\mathbf{a}$ but with its direction reversed. If k is negative, then $k\,\mathbf{a} = (-k) \cdot (-\mathbf{a})$.

If $\mathbf{a}$ is a velocity of a particle and $\mathbf{b}$ is its velocity at a later time, then $\mathbf{b} - \mathbf{a}$ is the increase in velocity over that time interval. If the length of the time interval is Δt, a vector with the same direction as $\mathbf{b} - \mathbf{a}$ and with a magnitude equal to the magnitude of $\mathbf{b} - \mathbf{a}$ divided by the scalar Δt is the *average acceleration* of the particle over the time interval Δt.

NOTES

Chapter 1: The Survival of the Valid

1. Some scholars object that the first known mention of this mandate of Plato was actually by Simplicius in the sixth century CE.
2. See Van der Waerden, 1975.

Chapter 2: The Bowl of Night

1. See Kuhn (1985), chapters 1 and 3.
2. Item 1 was noted by the first-century Greek geographer Strabo (1917, Vol. I, Bk. I para. 20). Items 2 and 3 are mentioned by Aristotle (1939, Bk. II, chap. 11).
3. My thesis advisor, Charles Loewner (1893–1968), grew up in a town near Prague. Loewner told of a large clock, shown in the figure, displayed outside a building in the Jewish quarter of Prague. The unusual thing about this clock is that its hands revolve counterclockwise. The Hebrew numerals, starting with aleph (א) representing 1, wind counterclockwise around the clock face. Loewner said that this clock inspired him to think whether there could be an abstract mathematical characterization of "clockwise." Photograph by Peter Benson, 2007.
4. This happens because Syene/Aswan at latitude 24° 4′ North is close to the boundary of the tropic of Cancer (currently latitude 23.5° North, but, due to the the precession of the obliquity of the ecliptic—the nutation of Earth's axis of rotation—about 1/4° larger in the time of Erastosthenes. The latitude of Alexandria is 31° 13′ North.
   There is a limit to this regularity. The rotation of the Earth slows somewhat irregularly due, for example, to the tides.
6. 180 degrees = π radians.
7. 60 minutes = 1 degree; 60 seconds = 1 minute.
8. One arcminute measured along a meridian of longitude on the surface of the Earth defines a nautical mile (= 1.15077945 statute miles). One knot is the unit of speed equal to 1 nautical mile per hour.
9. It would be more correct to say that they change very slowly over the centuries with the precession of the equinox because the vernal equinox makes a complete westward circle through the zodiac in 25,785 years.

10.   Over the centuries, the equation of time changes because the precession of equinoxes is not synchronous with the precession of the ecliptic.

11.   For more Information on the analemma, see the Internet site http://www.analemma.com.

12.   Figures 2.11a and b are celestial globes. The point P (e. g., Philadelphia) corresponds to a point on a concentric *geographical* globe whereas the analemma actually resides on the *celestial* globe.

13.   For the history of the analemma, see Sawyer 1994.

Chapter 3: Epicycles and Relative Motion

1.   *For readers familiar with complex numbers:* Put $\operatorname{cis} \theta = \cos \theta + i \sin \theta$. Then, as time t ranges over a suitable interval, an epicyclic curve, such as Figures 3.2 or 3.4, can be traced by a complex number of the form

$$r_1 \operatorname{cis} \omega_1 (t - t_0) + r_2 \operatorname{cis} \omega_2 (t - t_0)$$

where t_0 is a suitably chosen initial time.

2.   The labeled initial positions of the planets in these two diagrams are not intended to represent the configuration of the planets at a particular date.

3.   The tracing point on the epicycle travels with uniform circular motion relative to its center, which travels with uniform circular motion about the deferent circle.

Chapter 4: The Deferent–Epicycle Model

1.   Pluto has recently been reclassified, together with Xena and others, as a *dwarf planet*.

2.   *Datafitting* (also called *curvefitting*): any technique that approximates new values based on given data. In particular, *interpolation* or *extrapolation* methods create values that are "between" or "beyond" the given data values. Data-fitting methods generally have no bearing on the actual meaning of the data. "Curve" in curvefitting refers to the fact that sometimes the created data are visualized graphically, but, in general, no curve need be involved.

3.   The motion of the planets is periodic from the heliocentric—but not the geocentric—premise.

4.   An astronomical unit (AU) is the mean distance of Earth from the Sun, about 93 million miles. This distance was not known to Ptolemy.

5.   See Figure 4.2. An opposition occurs when Mars, Earth, and the Sun are aligned in that order—with Mars and the Sun on opposite sides of Earth.

6.   This characterization of the synodic period applies not only to Mars but also to the other *superior* planets. For the inferior planets (Venus and Mercury), the synodic period is the time between successive inferior conjunctions.

7.   The contrary assumption, $r_1 \geq r_2$, leads to the model in which deferent and epicycle are interchanged.

8.   S, S′, and S are not supposed to represent the actual *positions* of the Sun. This argument is an attempt to see a line of thought that would have been accessible to Ptolemy—who did not see how to locate the actual positions of the Sun.

9.   The terms *inside* and *outside* are used to avoid terms that have special planetary definitions, such as *inner* and *outer* or *inferior* and *superior*.

10.   Astronomers who are familiar with the *law of sines*—Ptolemy can be included in their number—could calculate the above result without waiting for the moment when EO′M′

is a right angle. Knowing the time elapsed from the most recent opposition, one can determine angle EO'M' from the knowledge of ω_1 and ω_2. Angle β can be determined by observation. Put $\gamma = $ angle EM'O'. Then, from the law of sines, $r_2 : r_1 = \sin \gamma : \sin \beta$.

11. The fraction $79/42$ arises in the continued fraction expansion of the length of Mars circuit of the Sun (1.8808476 years). For more on continued fractions, see Benson 1999, p. 245.

12. The theory of continued fractions provides a systematic method for approximating real numbers by rational numbers. See Benson 1999.

Chapter 6: The Reluctant Revolutionary

1. See Gingerich & MacLachlan, 2004, pp. 57–59.
2. http://www.maa.org/devlin/devlin_3_00.html
3. The following exercise in trigonometry might interest some readers: show that the shape of the phase of Venus determines the angles a and β. Furthermore, knowledge of either angle can be used to calculate the other angle and the shape of the crescent phase. Show the percentage of the disk of Venus that is illuminated is $50(1 - \cos(a + \beta))$.
4. From the geocentric point of view, Galileo's observations merely show that the epicycle of Venus must loop around the Sun.

Chapter 7: Circles No More

1. A locus is a mathematical term for a set of points satisfying a particular condition.
2. Let a planar curve C and a point P not on C be given. Let Q be the intersection of line $\mathcal{L}$ containing P and a tangent $\mathcal{T}$ to C such that $\mathcal{T}$ is perpendicular to $\mathcal{L}$. The locus of all such points Q is called the *pedal* curve of C with respect to point P. Proposition 7.3 says that the major circle is the pedal curve of an ellipse with respect to either of its foci.
3. Newton solves the problem of constructing an ellipse tangent to five given lines in *Principia*, Book 1, Proposition 27 (I. Newton, 1687).

Chapter 8: The War with Mars

1. More precisely, Kepler placed the Sun at one of the foci of each of the elliptical orbits of the planets.
2. Available now in English translation as *New Astronomy* (Kepler, 1992).
3. If p is a positive number, the base-10 logarithm of p is a number q such that $10^q = p$. Logarithms are useful for calculation because, for example, the logarithm of a product of two positive numbers is the sum of the logarithms of the factors: $\log ab = \log a + \log b$. To use a table of logarithms to compute the product ab, (1) look up $\log a$ and $\log b$, (2) add these two numbers, and (3) look up the number whose logarithm is equal to this sum. The slide rule procedure for computing the product is based on this computation.
4. Matthew 13:57.
5. Florence Nightingale (1820–1910) was a pioneer in the use of graphs; she graphed the incidence of disease in the Crimean war. See Benson, 1999.
6. For further discussion of the vibrating string, see Benson, 2003, ch. 3.
7. The velocity of an object is its speed *and* its direction of motion. This distinction between speed and velocity will be discussed later in Section 10.1.
8. A casual examination of diagrams in Kepler's *Astronomia nova* reveals that an angle of about $43°$ often appears—a clue that Kepler is using the parallax method described here.

Chapter 9: The Birth of Mechanics

1. See Benson, 1999, p. 198, for details.
2. However, it can be shown that the center of mass of a triangle consisting of three rods of uniform density is, in general, *not* the intersection of the medians.
3. *Block* is the nautical term for pulley, and *tackle* refers to the ropes of a ship's rigging.
4. Strictly speaking, since the weights move along a circular arc, δ_1 and δ_2 should be defined as the *vertical* components of the displacement. Nevertheless, the proportion 2 : 1 holds as claimed.
5. The more general Principle of Virtual Work speaks of *infinitesimal* virtual displacements— a calculus concept that was not clearly understood when first discussed and that was made rigorous only late in the nineteenth century.
6. See also Benson, 2003, pp. 177–183.
7. See also Benson, 2003, page 135.

Chapter 10: The Astronomical Alchemist

1. Infinitesimals made a comeback in the 1960s in the *nonstandard analysis* of Abraham Robinson.
2. A recent English translation (I. Newton, 1687) is available.
3. For a particle with mass m and velocity v, Leibniz defined its *vis viva* to be mv^2. Today, we would define its *kinetic energy* to be half of that quantity.
4. The uppercase Greek letter Δ is often used to denote an *increment* of a variable: $\Delta\theta$, Δt, ΔA, etc.
5. Exceptions occur in the unlikely case that the ball is thrown faster than the speed imparted to the ball by Earth's rotation.
6. Art is only approximately an inertial observer because he participates in the rotation of the entire planet, as well as the movement of Earth in the motion around the Sun. It can be shown that the effect of these rotational movements are negligible in the example under discussion.
7. The angle α of rotation can be computed as follows. In radian measure, the value of this angle is the ratio of $\overline{PU}$ to the radius of $\mathcal{N}$, which is equal to $\sqrt{3} = 99.239°$.
8. McIntyre (2000) discusses many examples of Coriolis motion, including the one shown in Figure 10.7.
9. We assume that the shove does not exceed the orbital velocity.
10. The proof uses the fact that $\cos 60°$ is equal to $1/2$.
11. The *inverse* problem was so-named in the eighteenth century. The modern preference would be to call it the *converse* problem.
12. In his proof of Proposition 70 in Book 1 of *Principia*, Newton tacitly assumes that C is an equator of S, which is true only for special choices of points P, Q, and R. This seems to be a gap in Newton's proof because it is necessary to determine how projection distorts distances for *arbitrary* choices of points P, Q, and R.
13. The times for the journeys of Photons One and Two are, respectively,

$$\frac{2w}{\sqrt{c^2 - v^2}} \quad \text{and} \quad \frac{w}{c + v} + \frac{w}{c - v}$$

For example, if c and v are, respectively, 0.5 and 0.3 miles per hour, and $w = 1$ mile, then the times for Photons One and Two are, respectively,

$$\frac{2}{0.4} = 5 \text{ hours, and } \frac{1}{0.8} + \frac{1}{0.2} = 1.25 + 5.00 = 6.25 \text{ hours}$$

That Photon One always wins is a consequence of the inequality of the arithmetic and geometric means, which asserts that, for arbitrary nonnegative numbers a and b, $\sqrt{ab} \leq (a + b)/2$, where equality holds only if $a = b$.

14. Surprisingly, Einstein's creation of the theory of relativity may have been motivated by a problem in electrodynamics rather than the Michelson-Morley experiments (see Holton, 1973, chapter 8). Nevertheless, Michelson-Morley is the logical precursor of relativity and the simpler path of introduction to relativity.

REFERENCES

Aristotle. (1939). *On the heavens.* Harvard University Press, Loeb Classical Library.

Benson, D. C. (1999). *The moment of proof: Mathematical epiphanies.* New York: Oxford University Press.

Benson, D. C. (2003). *A smoother pebble: Mathematical explorations.* New York: Oxford University Press.

Brown, D. (2003). *The Da Vinci code.* New York: Doubleday.

Gingerich, O. (1993). *The eye of heaven.* New York: American Institute of Physics.

Gingerich, O., & MacLachlan, J. (2004). *Nicolaus Copernicus: Making the earth a planet.* New York: Oxford University Press.

Goodstein, D. L., & Goodstein, J. R. (1996). *Feynman's lost lecture.* New York: Norton.

Hamilton, W. R. (1847). The hodograph, or a new method of expressing in symbolical language the Newtonian law of attraction. *Proceedings of the Royal Irish Academy,* 3, 344–353. (Edited by David R.Wilkins, http://www.maths.tcd.ie/pub/HistMath/People/Hamilton/Hodograph/)

Holton, G. (1973). *Thematic origins of scientific thought.* Cambridge, Mass.: Harvard University Press.

Kepler, J. (1992). *New astronomy.* New York: Cambridge University Press. (Translated by-William H. Donahue. First published in 1609 in Latin under the title *Astronomia nova.*)

Kuhn, T. S. (1985). *The Copernican revolution.* Cambridge, Mass: Harvard University Press.

McIntyre, D. H. (2000). Using great circles to understand motion on a rotating sphere. *American Journal of Physics,* 68(12), 1097–1105.

Newton, I. (1687). *The Principia.* Berkeley: University of California Press. (Republished in 1999. Translation from Latin and commentary by I. Bernard Cohen and Anne Whitman, assisted by Julia Budenz. Including *Guide to Newtons Principia* by I. Bernard Cohen.)

Newton, R. R. (1977). *The crime of Claudius Ptolemy.* Baltimore, Md.: Johns Hopkins University Press.

Ptolemy. (1998). *Ptolemy's Almagest.* Princeton, N.J.: Princeton University Press. (Translated and annotated by G.J. Toomer.)

Sawyer, F. W. I. (1994). Of analemmas, mean time, and the analemmatic sundial, http://www.longwoodgardens.org/Sundial/Analemma.html. *Bulletin of the British Sundial Society.*

Strabo. (1917). *The geography of strabo.* Loeb Classical Library.

Van der Waerden, B. L. (1975). *Science awakening* (Vol. 1). Groningen: P. Noordhoff. (Originally published by P. Noordhoff in 1961 as *Ontwakende wetenschap.*)

Weinstock, R. (2000). Inverse-square orbits in Newton's Principia. *Arch. Hist. Exact Sci.,* 55, 137–162.

INDEX